Introduction to English Morphology

Textbooks in English Language and Linguistics (TELL)

Edited by Magnus Huber and Joybrato Mukherjee

Volume 5

PETER LANG
Frankfurt am Main · Berlin · Bern · Bruxelles · New York · Oxford · Wien

Alexander Tokar

Introduction to English Morphology

PETER LANG
Internationaler Verlag der Wissenschaften

Bibliographic Information published by the Deutsche Nationalbibliothek
The Deutsche Nationalbibliothek lists this publication in the Deutsche Nationalbibliografie; detailed bibliographic data is available in the internet at http://dnb.d-nb.de.

Cover Design:
© Olaf Gloeckler, Atelier Platen, Friedberg

ISSN 1862-510X
ISBN 978-3-631-61841-7
© Peter Lang GmbH
Internationaler Verlag der Wissenschaften
Frankfurt am Main 2012
All rights reserved.

All parts of this publication are protected by copyright. Any utilisation outside the strict limits of the copyright law, without the permission of the publisher, is forbidden and liable to prosecution. This applies in particular to reproductions, translations, microfilming, and storage and processing in electronic retrieval systems.

www.peterlang.de

To Dieter Stein

Preface

This textbook grew out of the seminars and lectures on various aspects of English theoretical linguistics (e.g. morphology, morphosyntax, lexicology, phraseology, phonetics and phonology, etc.) which I taught at the Universities of Düsseldorf and Gießen between April 2006 and February 2011.

Like the previous volumes in the TELL Series, this introduction to English morphology is intended as a companion for students of English language and linguistics throughout their studies. This means that the book contains a discussion of both 1) very basic introductory issues requiring no or very little prior background in linguistics (e.g. what is a morpheme?) and 2) fairly controversial theoretical issues such as, for example, the question of whether a fully-idiomatic word like *understand* can be segmented into the morphemes {under} and {stand}, to which different linguists provide different answers. The textbook can thus be used by students at both B.A. and M.A. levels.

The book consists of six chapters. Chapter 1 is an introductory chapter that provides a definition of morphology as a branch of linguistics. Chapter 2 dwells on the internal structure of English words. Chapter 3 discusses the formal and semantic structure of English lexemes. Chapters 4 and 5 are concerned with word-formation in English: special emphasis is laid on the methodological question of how students of English morphology can determine whether a particular word-formation mechanism is still productive in Present-day English. Finally, Chapter 6 discusses grammatical categories in English such as tense, aspect, voice, number, case, etc. Each of the chapters ends with exercises and suggestions for further reading.

The present textbook departs from other recent introductions to morphology in the following important respects:

- Inspired by the recent article Haspelmath (2011), it rejects the orthographic criterion for distinguishing between free and bound forms.

- Following Mel'čuk (1968), it defines a linguistic sign not as a doublet consisting of the signifier and the signified (as defined by Ferdinand de Saussure), but as a conventionalized association between the signifier, the signified, the syntactics, and the sociolinguistics.

- It classifies word-formation into lexeme-formation and lex-formation.

Like the authors of the previous TELL volumes, I have attempted to illustrate my analyses with authentic language data, i.e. examples drawn from linguistic corpora such as the Corpus of Contemporary American English and British National Corpus. In addition, I have extensively consulted the Oxford English Dictionary and Paul McFedries's online-based collection of English neologisms Word Spy (http://www.wordspy.com/).

I would like to thank Professor Magnus Huber for inviting me to write this volume for the TELL Series and for providing extremely helpful comments on an earlier version of this manuscript. I am also deeply indebted to Wiebke Ostermann and Louise Nieroba for their scrupulous proof-reading of the final version of the manuscript.

This textbook has undoubtedly benefited from my conversation with Professor Vladimir Plungian in February 2010 in Moscow. Vladimir Plungian is the author of *Introduction to General Morphology* (Plungian 2000), the textbook which has sparked off my own interest to morphology and was a great source of inspiration for me when I was writing this book.

Last but not least, I wish to thank Kate Butkus, Sergey Danilov, Patrick Maiwald, Bridgit Nelezen, and Stefan Storm for insightful discussions and many valuable comments on some theoretical issues relating to this volume.

I alone am responsible for any remaining errors and shortcomings.

This book is dedicated to Professor Dieter Stein. I would like to congratulate him on his 65th birthday and wish him many more years of good health and successful linguistic research.

Düsseldorf   Alexander Tokar
November 2011

Table of Contents

1 Basic concepts	**1**
1.1 What is morphology?	1
1.2 Morphology and other branches of linguistics	3
1.2.1 Semantics	3
1.2.2 Phraseology	5
1.2.3 Phonetics and phonology	7
1.2.4 Syntax	9
1.2.5 Sociolinguistics	12
1.3 What is a word?	14
1.4 Exercises	21
1.5 Further reading	22
2 The internal structure of English words	**25**
2.1 What is a morpheme?	25
2.2 Morphemes as signs	27
2.2.1 One signifier → more than one signified	28
2.2.2 One signified ← more than one signifier	28
2.2.3 The syntactics of a sign	29
2.2.4 The sociolinguistics of a sign	30
2.2.5 The signified as the most important sign component	31
2.3 The distribution of morphs	33
2.4 The segmentation of words into morphemes	39
2.4.1 Anisomorphism. Full-idiomaticity	39
2.4.2 A purely semantic approach	40
2.4.3 Nida's purely formal approach	40
2.4.4 Nida's approach and the conception of differential meaning	43
2.4.5 Mel'čuk's theory of quasi-linguistic units	44
2.4.6 Anisomorphism. Partial idiomaticity	45
2.4.7 Anisomorphism. Additional meanings	47
2.5 The hierarchy of morphs and units alike	50
2.5.1 Affixes versus roots	51
2.5.2 Combining form as a distinct morpheme type?	53
2.5.3 One signifier → both a root and an affix	55
2.5.4 Typology of affixes	56

	2.5.5 Typology of roots	59
	2.6 Exercises	60
	2.7 Further reading	61
3	**Analyzing English lexemes**	**63**
	3.1 What is a lexeme?	63
	3.2 The structure of a lexeme	64
	3.2.1 The lex of a lexeme	64
	3.2.2 The typology of lexes	66
	3.2.3 The signified of a lexeme	66
	3.2.4 Three-component anisomorphic lexemes	67
	3.2.5 Anisomorphic lexemes realized by phrases and sentences	68
	3.2.6 How to distinguish between full-, semi-, and quasi-idioms?	70
	3.3 Lexemes and vocables	72
	3.3.1 Relations between members of the same vocable	72
	3.4 Lexemes and lexeme families	74
	3.5 Exercises	75
	3.6 Further reading	76
4	**Word-formation: basic issues**	**79**
	4.1 Lexeme-formation versus lex-formation	79
	4.2 Lexeme-formation	80
	4.2.1 Purely semantic mechanisms	80
	4.2.2 Purely formal mechanisms	81
	4.2.3 Mechanisms involving formal and semantic modifications	83
	4.2.4 Diachronic and synchronic perspectives	86
	4.2.5 Why do speakers of English create new lexemes?	89
	4.2.6 The establishment of new lexemes	92
	4.2.7 The non-institutionalization of new lexemes	94
	4.2.8 Productivity	99
	4.3 Lex-formation	103
	4.3.1 Lex-forming clipping	104
	4.3.2 Lex-forming suppletion	104
	4.3.3 Lex-forming abbreviation	105
	4.3.4 Lex-forming borrowing	106
	4.3.5 Lex-forming apophony	106
	4.3.6 Lex-forming affixation	106
	4.3.7 Lex-forming syntactics' change	107
	4.3.8 Lex-forming orthographic modification	108

4.4 Exercises	108
4.5 Further reading	110
5 Lexeme-building mechanisms	**111**
5.1 Semantic change	111
5.1.1 Mechanisms of semantic change	112
5.1.2 Types of metonymies	112
5.1.3 Types of metaphors	114
5.1.4 Morphological conversion	115
5.1.5 Productivity	124
5.2 Lexeme-manufacturing	127
5.2.1 Productivity	129
5.3 Lexeme-building borrowing	130
5.3.1 Productivity	131
5.4 Lexeme-building affixation	133
5.4.1 Affixes and their signifieds	133
5.4.2 Affixes and their syntactics	134
5.4.3 Productivity	141
5.5 Lexeme-building apophony	143
5.5.1 Productivity	145
5.6 Compounding	145
5.6.1 Compounding as an anisomorphic mechanism	146
5.6.2 The semantics of compounding	148
5.6.3 Endocentric and exocentric compounding	152
5.6.4 Compounding from a formal point of view	156
5.6.5 Compounds and phrases	159
5.6.6 Productivity	162
5.7 Blending	163
5.7.1 Productivity	165
5.8 Idiomatization of phrases and sentences	166
5.8.1 Productivity	167
5.9 Back-formation	168
5.9.1 Productivity	170
5.10 Exercises	171
5.11 Further reading	172
6 Inflectional morphology	**173**
6.1 Grammatical category	173
6.2 Types of grammatical categories	174

6.3 Wordform-building mechanisms	176
6.3.1 Inflectional affixation	176
6.3.2 Analytic formation	177
6.3.3 Grammatical apophony	177
6.3.4 Grammatical suppletion	180
6.3.5 Signifier-sharing	181
6.3.6 Allowordforms	181
6.3.7 Productivity	182
6.4 Syntactic grammemes in English	183
6.4.1 Why do we need the passive voice?	183
6.4.2 Middle voice in English?	184
6.4.3 What can be passivized?	186
6.4.4 *Get*-passives	187
6.4.5 Types of cases	188
6.4.6 Cases in English	189
6.4.7 Functions of case	191
6.4.8 Semantic function of case?	192
6.5 Semantic grammemes in English	194
6.5.1 Typology of temporal meanings	194
6.5.2 No future tense in English	196
6.5.3 Idiomatic uses of temporal wordforms	198
6.5.4 Typology of aspectual meanings	200
6.5.5 Aspects in English	202
6.5.6 Moods in English	205
6.5.7 Person	209
6.5.8 Number	212
6.5.9 Degrees of comparison	214
6.5.10 Numerical qualification	216
6.6 Exercises	218
6.7 Further reading	220
Key to exercises	**221**
References	**225**
Index	**233**

1   Basic concepts

It has become a tradition to begin monographs and textbooks on morphology with a tribute to the German poet Johann Wolfgang von Goethe, who invented the term *Morphologie* in 1790 to refer to "eine Anschauung von den Gestalten und Wandlungen der Natur und Kunst" / 'a view about the forms and transformations of nature and art' (Kluge 2002). In the English language the word *morphology* has been used since 1828: according to the Oxford English Dictionary (henceforth OED), it originally referred to "the branch of biology that deals with the form of living organisms and their parts, and the relationships between their structures".

In this introductory chapter we will learn in which respects our present-day linguistic understanding of the term 'morphology' differs from that of Goethe and that of the 19th century biologists. In addition, the chapter discusses the relationship between morphology and other branches of linguistics as well as one of the most important theoretical concepts in morphology: the concept of a word. With regard to the latter, our focus will be on the question of whether there is at least one operational criterion, or to be more precise, a formal feature characteristic of a particular combination of sounds justifying the treatment of that combination as a word. For example, why is it that the combination of the sounds /kæt/ is usually regarded as a word? What is the difference between the word *cat* and, say, *black cat*, which is usually regarded as a combination of no less than two words: *black* and *cat*?

1.1   What is morphology?

In a linguistic context morphology is usually defined as the study of the internal structure of words. To illustrate what this means, let us consider the word *waithood* 'the stage in a young college graduate's life when activities such as marrying and finding a place to live are postponed until a job is found or enough money is saved'. According to Word Spy (i.e. a freely available online database of English neologisms which can be accessed at http://www.wordspy.com/), *waithood* is a relatively new word in English: its earliest citation provided by Word Spy dates September 01, 2007.

From a formal point of view, *waithood* seems to be segmentable into the components *wait* and *-hood*. The component *wait* is a verb which, apart from occurring in *waithood*, also occurs in sentences like *I cannot wait any longer*. And the component *-hood* is a noun-building element which, apart from occurring in *waithood*, also occurs in nouns like *adulthood, boyhood,* *parenthood*, etc. Taking this into account, we can draw an important conclusion, as far as the internal structure of *waithood* is concerned. The word under analysis is a **complex word**, i.e. a word which can be analyzed as a combination of no less than two formally identifiable components capable of occurring in other morphologically relevant environments. (In the next section of this chapter, we will learn which environments qualify as morphologically relevant.) By contrast, the combination of the sounds /kæt/ representing the word *cat* cannot be segmented into /k/ and /æt/, or /kæ/ and /t/, or /k/ and /æ/ and /t/. Undeniably, these putative components occur elsewhere (e.g. *cable*, *at*, *cap*, *mat*), but none of these environments seem to be morphologically relevant. Accordingly, the word *cat* is a **simple word** (or a **simplex**), i.e. a word which consists of no more than one formally indivisible component.

In addition to determining whether the word under analysis is a simple or complex word, morphology also deals with the question of how new words like the above mentioned *waithood* come into existence. Compare, for example, the formation of *waithood* and that of the verb *to wife* 'to downplay a woman's career accomplishments in favor of her abilities as wife and mother' (Word Spy). As we established above, *waithood* is, from a formal point of view, segmentable into the components *wait* and *-hood*. Accordingly, we can conclude that the word under analysis was created via combining the components *wait* and *-hood*: *waithood* = *wait* + *-hood*. By contrast, the creation of the formally simple verb *to wife* seems to have involved a **semantic modification** of the already existing noun *wife*. Thus the meaning 'to wife' is **semantically more complex** than the meaning 'a wife': the former contains the meaning 'a wife' plus the additional meaning 'to downplay a woman's career accomplishments in favor of her abilities as mother'. Accordingly, we can conclude that the meaning 'a wife' served as an **input meaning** for the meaning 'to wife'.

Finally, morphology studies those modifications (of existing words) that do not give rise to new words but serve to express **grammatical meanings** such as, for example, 'plurality', 'the past tense', 'the passive voice', etc. For instance, the addition of *-s* to the noun *book* does not create a new word: both *book* and *books* refer to representatives of the same class of objects (books). However, *book* and *books* differ with regard to their grammatical meanings: while *book* is in the **singular number**, *books* is in the **plural number**. The singular–plural opposition exemplified by *book* and *books* forms the **grammatical category** NUMBER. A similar case is the **present tense–past tense** opposition, which forms the grammatical category TENSE: both *work* of *I work* and *worked* of *I worked* can refer to the same action of working. However, the forms under consideration differ with regard to their grammatical meanings: while *work* is in the present tense, *worked* is in the past tense.

The present textbook contains a detailed discussion of each of the three aspects of morphology named above. Chapters 2 and 3 deal with both the formal and the semantic structure of English words: among other things, we will become acquainted with the types of components into which complex words like *waithood* can be segmented. Chapters 4 and 5 deal with what has usually been called **word-formation**, i.e. processes that produce new words like *waithood* and *to wife*. Finally, Chapter 6 deals with grammatical categories in English such as NUMBER, TENSE, VOICE, etc.

1.2 Morphology and other branches of linguistics

To be able to address the aforementioned issues, we will, first of all, need to become acquainted with the most important theoretical concepts pertaining to other branches of linguistics which morphology is closely connected to. These include semantics, phraseology, phonetics and phonology, syntax, and sociolinguistics.

1.2.1 Semantics

Semantics is the branch of linguistics which is concerned with meaning. To demonstrate in what ways morphology is dependent on semantics, let us again compare the internal structure of the words *waithood* and *cat*.

The reason why the occurrence of *wait* in sentences like *I cannot wait any longer* and that of *-hood* in nouns like *adulthood, boyhood, parenthood*, etc. can be considered morphologically relevant is that the overall meaning 'waithood' is closely connected to the meanings 'to wait' and 'stage', which the components *wait* and *-hood* express in other environments: e.g. while adulthood can be defined as the stage in our lives that involves being an adult (or, alternatively, the state of being an adult), waithood can be defined as a stage in the life of a college graduate that involves waiting. This fact is the main justification for the formal segmentation of *waithood* into the components *wait* and *-hood*.

In contrast, there is no similar justification for the segmentation of *cat* into /k/ and /æt/, or /kæ/ and /t/, or /k/ and /æ/ and /t/. In neither *cat* nor words like *cable, at, cap, mat*, etc., where these sounds occur as well, do they express discernible meanings of their own. Thus the meaning 'cat' does not seem to be a complex meaning segmentable into at least two independent meanings – e.g. the meanings 'animal' and 'characteristic features of cats distinguishing them from other animals' – which one could attribute to the putative components /k/ and /æt/ or /kæ/ and /t/. What justifies this claim is that other words containing these sounds (or sequences of these sounds) do not express related meanings. For example, neither the meaning of *cable*, which contains the sound [k], nor the meaning of *at*, which contains the sound sequence [æt], seem to have anything in common with the meaning 'cat'. The same is true of the word *cap*, which contains the sound sequence [kæ], and the word *mat*, which contains the sound [t]: neither the former nor the latter are semantically related to *cat*.

In summary, a morphological analysis of the internal structure of a word depends on the semantic analysis of the meaning of that word. Accordingly, a morphologist analyzing the internal structure of some word must have a very clear idea of what that word means. But what precisely is meant by *the meaning of a word*? What does it mean *to mean something*? How do we know that e.g. the word *waithood* means 'a particular stage in the life of a college graduate that involves waiting'?

In semantics meaning is usually defined as the **concept** associated with a particular sound form. A concept is a mental description activated by that sound form. For instance, those speakers of English who are familiar with the word *waithood* have the concept WAITHOOD stored in their **mental lexica**. That is, as soon as they hear the sound form /ˈweɪthʊd/, their minds 'picture' a college graduate who postpones marrying until he or she finds a good job or saves enough money.

The mental lexicon is often described as a kind of dictionary (which we have in our brains) consisting of multiple sound form–concept correspondences like the one exemplified by *waithood*. That is, the mental lexicon of the average speaker of English contains the correspondences between e.g. the sound form /kæt/ and the concept CAT (i.e. a description of an animal called *cat*) associated with it; the sound form /ˈtiːtʃə(r)/ and the concept TEACHER (a description of a person who teaches); the sound form /dɪˈmɒkrəsɪ/ and the concept DEMOCRACY (a description of a democratic political system); etc.

The key task of semantics is thus the description of the concept constituting the meaning of a particular linguistic expression. For this purpose, a semanticist can resort to two strategies. One is the identification of so-called **necessary and sufficient conditions**. These are the 'minimum requirements' whose fulfillment suffices to qualify as a member of a particular **conceptual category**. For example, it is often argued that the minimum requirement that is fulfilled by all representatives of the conceptual category MOTHER is that of being a female parent: any entity who is both female and a parent can be referred to as someone's mother. Accordingly, the features [female] and [parent] can be said to constitute the meaning of the word *mother*.

An alternative to this is the **prototype approach**. Its essence is the identification of characteristics applying to the best representative of a given conceptual category, its **prototype**. A prototypical mother, for instance, is a person

> who is and always has been female, and who gave birth to the child, supplied her half of the child's genes, nurtured the child, is married to

the father, is one generation older than the child, and is the child's legal guardian. (Lakoff 1987: 83)

One of the most controversial issues in semantics is the question of whether linguistic meaning can be equated with **encyclopedic knowledge**. That is, for example, can we really describe the meaning of *mother* in terms of the feature [supplied her half of the child's genes], given that a number of people – e.g. small children who have not yet had a biology class – are not really aware of this fact but nevertheless can easily distinguish between their own mothers and other females?

According to Wierzbicka (1972: 46), *mother* means 'the human being inside whose body (once) there was something that was becoming another person's body'. (The word *father* then means 'the human being who once caused a woman to have inside her body something that was becoming another person's body'.) Undeniably, this is a much less encyclopedic definition than that which makes use of the feature [supplied her half of the child's genes]. However, it is not non-encyclopedic: the fact that a mother is a person who inside her body once used to have the body of another person is also an encyclopedic fact that may not be shared by all members of the English linguistic community.

1.2.2 Phraseology

Phraseology is the branch of semantics which is concerned with **idiomatic meanings** (in particular, with combinations of more than one word such as e.g. *kick the bucket* which have idiomatic meanings). According to Dobrovol'skij and Piirainen (2005: 40), there are two types of idiomatic meanings: those that involve **semantic modification** or **opacity** of the literal meaning of the form under analysis. (The term '**literal meaning**' can be defined negatively: a literal meaning is neither an opaque nor a modified meaning.)

The difference between the two types of idiomatic meanings is as follows. In the case of semantic modification, the average speaker (i.e. a layman lacking any expertise in semantics and phraseology) can easily explain to him- or herself why the idiomatic word in question does not mean what its components literally stand for. By contrast, in the case of opacity, the idiomatic meaning is not transparent, i.e. cannot be accounted for in terms of the components' literal meanings.

As an illustration of a modified meaning, let us consider the meaning of *boyfriend* in (1).

(1)   *She's decided to break up with her <u>boyfriend</u>* (Corpus of Contemporary American English, henceforth COCA)

From a formal point of view, the word *boyfriend* seems to be segmentable into the components *boy* and *friend*. However, the meaning 'boyfriend' cannot be segmented into the meanings 'boy' and 'friend'. First of all, a boyfriend is not a boy (i.e. a young male child, which *boy* literally means) but a male of almost any age over puberty (Holder 2008: 104). Second, a boyfriend is not literally a friend (i.e. any person whom a woman 'know[s] well and regard[s] with affection and trust', as defined by WordNet) but a regular sexual partner in a non-marital sexual relationship. The meaning 'boyfriend' is thus a fully-idiomatic meaning that does not contain the meanings 'boy' and 'friend', inherent in the components *boy* and *friend* in other environments.

At the same time, however, notice that the literal meanings 'boy' and 'friend' partially **motivate** the idiomatic meaning 'boyfriend': boyfriends are often perceived and explicitly referred to as friends in the literal meaning of this word; e.g. in (2).

(2)   My *boyfriend* is my best *friend* (http://tinyurl.com/6ys8pzz)

Similarly, the idiomatic meaning 'a male of almost any age over puberty' is related to the meaning 'a young male child' in that both the former and the latter share the semantic component [male]: both a boyfriend and a boy are males. Because of these two facts, the average speaker of English can easily explain to him- or herself what boyfriends have in common with friends who are boys.

As an illustration of an opaque meaning, let us consider the meaning of the verb *understand*. Like *boyfriend*, this word seems to be formally segmentable into two components: *under* and *stand*. However, as in the case of *boyfriend*, the meaning 'to understand' does not contain the meanings 'under' and 'to stand': *understand* does not mean 'to stand under somebody or something' but 'to grasp the meaning or the reasonableness of something' (Merriam-Webster Online, henceforth MWO). But while a boyfriend is typically a friend and a former male child, understanding does not seem to have much in common with physical standing: standing under somebody or something does not necessarily result in the understanding of the person or the thing (under which you are standing).

Apart from fully-idiomatic words like *boyfriend* and *understand*, there are 1) partially idiomatic words like *blackboard*, whose overall meanings contain one of their components' literal meanings and 2) words like *waithood* whose overall meanings do not only contain their components' literal meanings but also some additional, unpredictable idiomatic meanings.

With regard to *blackboard*, it is obvious that its overall meaning contains the meaning of the component *board* but not of the component *black*. A blackboard is not a black board but a board for drawing or writing upon with chalk (MWO). Blackboards typically have dark surfaces (and are in this respect different from whiteboards[1]), but they are not always black: green blackboards, for instance, do occur as well.

With regard to *waithood*, observe that its meaning does not only contain the meanings 'to wait' and 'stage', inherent in the components *wait* and *-hood*: *waithood* does not mean 'any stage in the life of any person that has something to do with waiting' but 'a particular waiting stage: the one that involves a young college graduate who is waiting for financial security in his or her life and therefore postpones marrying'. The meaning 'waithood' is thus **narrower** than that of the mere sum of the meanings 'to wait' and 'stage', inherent in the components *wait* and *-hood*.

Idiomatic words like *boyfriend* and *understand* pose a particular theoretical challenge for a morphologist analyzing their internal structure. Can *boyfriend* and *understand* be regarded as complex words, even if *boyfriend* does not mean 'a friend who is a boy' and *understand* does not mean 'to stand under somebody or something'? This question will be addressed in Section 2.4 of the next chapter.

1.2.3 Phonetics and phonology

Phonetics and phonology are related linguistic disciplines which are both concerned with sound. Phonetics studies any physical aspect of sound. For example, how do we produce and perceive sounds? (The branch of phonetics which studies the production of sounds is called **articulatory phonetics**; the branch of phonetics which deals with the perception of sounds is known as **acoustic phonetics**.) Phonology, by contrast, is concerned with that aspect of sound that performs a specific linguistic function (Trubetzkoy 1969: 11). For example, the sound [k] of *cable* makes this word distinguishable from *table*. Similarly, the sound [ð] of *that* makes this word distinguishable from *chat*.

The **meaning-distinguishing function** exemplified by [k] of *cable* and [ð] of *that* is the most important but not the sole function of sound. According to Trubetzkoy (1969: 16), sounds can also perform the **expressive function**. This means that sounds we produce characterize us as members of particular (regional, social, etc.) groups. For example, **General American** (i.e. the standard accent of American English) differs from **Received Pronunciation** (the standard accent of British English) with regard to the pronunciation of the word *metal*. In General American this word is almost indistinguishable from

---

[1] Many English speakers are reluctant to use *blackboard* as a synonym of *whiteboard*. Accordingly, it can be argued that the former means 'a dark board used for writing or drawing upon with chalk', i.e. the meaning 'dark' is part of the overall meaning 'blackboard'. In this connection, it is important to emphasize that this fact does not undermine the analysis of *blackboard* as a partially idiomatic word: the meaning 'dark' is not the literal meaning of the component *black*.

medal: both words have the pronunciation /ˈmɛdl/.[2] In contrast, in Received Pronunciation *metal* is pronounced /ˈmɛtl̩/ and *medal* is pronounced /ˈmɛdl/. Accordingly, we can say that if a native speaker of English expresses the meaning 'metal' with the sound form /ˈmɛdl/ rather than /ˈmɛtl̩/, this person is (most likely) an American who has a General American accent. The sound [d] of *medal* can thus be said to perform the expressive function of characterizing those English speakers who pronounce *metal* /ˈmɛdl/ as members of the group 'General American speakers'.

The sounds [k] and [t] of *cable* and *table*, which make these two words distinguishable from each other, can be said to realize two different **phonemes**. A phoneme is usually defined as the smallest meaning-distinguishing linguistic unit. A sound that realizes a given phoneme is its **phone**. If one and the same phoneme is realized by more than one sound, we speak of **allophones**. Compare, for example, the articulation of the *t*-sounds in *take* and *eighth*. In *take*, it has an **alveolar articulation**, i.e. one which involves the raising of the tip of the tongue (its **active articulator**) towards the upper alveolar ridge (**passive articulator**). In *eighth*, the articulation is **dental**, i.e. the tip of the tongue is raised not towards the alveolar ridge but towards the upper teeth. The alveolar [t] of *take* and the dental [t̪] of *eighth* can be regarded as allophones of the same phoneme because these two sounds cannot perform the meaning-distinguishing function: if some speaker of English pronounces *take* with the dental [t̪] rather than with the alveolar [t], this will not give rise to a new word, i.e. /t̪eɪk/ will also be understood as a sound form representing the word *take*.

Allophones of a phoneme can either be in **complementary distribution** or in **free variation**. The former involves two or more sounds that occur in different environments, thereby complementing each other. For example, the dental [t̪] occurs before the interdental sounds [θ] and [ð], but not before vowels; in the latter environment the preference is usually given to the alveolar [t]. As regards free variation, consider the word *economic*. As McMahon (2002: 58) points out, one and the same speaker of English can pronounce this word as /ˌɛkəˈnɒmɪk/ on one occasion and as /ˌiːkəˈnɒmɪk/ on another occasion. The initial vowel sounds [ɛ] and [iː] can thus be regarded as allophones of the same phoneme which occur in free variation: these sounds are interchangeable in the same environment without a change in meaning.

Phones that realize two different phonemes – for example, [k] of *cable* and [t] of *table* – are in **contrastive distribution**. In connection with contrastive distribution, notice that two sounds that are in free variation in one environment can be in contrastive distribution in another environment. For instance, the sounds [ɛ] and [iː], which are in free variation in *economic*, perform the meaning-distinguishing function in the words *hell* /hɛl/ and *heal* /hiːl/.

---

[2] Most pronunciation transcriptions used in this book are taken from the OED.

As we will see in Chapters 2 and 3, the categories of contrastive distribution, complementary distribution, and free variation elaborated on above can also be applied at the morphological level, describing the distribution of words and their meaningful components.

1.2.4 Syntax

Syntax is a branch of linguistics which is concerned with units larger than the word. These include:

1. **the phrase**
2. **the clause**
3. **the sentence**

A phrase is either an individual word or (more often) a combination of words capable of performing a **syntactic function**. The most important syntactic functions are the **subject** and the **predicate**. Consider, for example, (3).

(3)   *The French are fighting in Afghanistan* (COCA)

This sentence can be segmented into two phrases: *the French*, functioning as subject, and *are fighting in Afghanistan*, functioning as predicate.

Phrases like *the French* of (3) functioning as subjects have a number of formal properties that make them distinguishable from phrases performing other syntactic functions. According to Huddleston (2002a: 236-237), these properties are as follows.

- The default position of the subject in English is before the **main verb** (i.e. a verb like *fight* of (3) that has a lexical meaning). Cf. *The French are fighting in Afghanistan* and **In Afghanistan are fighting the French*.

- In **interrogative sentences** the subject follows the **auxiliary verb** (i.e. a verb like *are* of (3) that helps to express a grammatical meaning). Cf. *Are the French fighting in Afghanistan?* and **Are in Afghanistan the French fighting?*

- The subject **agrees** with the verb in **person** and number. Cf. *The French are fighting in Afghanistan* and **The French am fighting in Afghanistan*.

- The subject position can be filled by a pronoun in the **nominative**, but not in the **accusative** or the **genitive case**. Cf. *They are fighting in Afghanistan* and **Them are fighting in Afghanistan* or **Their are fighting in Afghanistan*.

- The subject of a **declarative clause** agrees in person and number with the subject pronoun of an appended **interrogative tag**. Cf. *The French are fighting in Afghanistan, aren't they?* and **The French are fighting in Afghanistan, isn't he?*

- The subject is generally an obligatory element. Cf. *The French are fighting in Afghanistan* and **Are fighting in Afghanistan*.

There are no similar formal criteria justifying the isolation of the predicate. However, the predicate can be isolated negatively. The predicate is that part of a sentence like (3) which does not qualify as the subject. In (3), for example, the phrase *are fighting in Afghanistan* qualifies as the predicate simply because it is not part of the subject *the French*.

The predicate usually consists of the **predicator** and the **predicative**. The predicator is the verb which **heads** the predicate phrase; e.g. *are fighting* of *are fighting in Afghanistan*. The predicative is the rest of the predicate phrase; e.g. *in Afghanistan* of *are fighting in Afghanistan*. (There may be predicates consisting of the predicator only. For example, *are fighting* of *The French are fighting*.)

Predicatives can be classified into:

1. **objects**
2. **complements**
3. **adjuncts**

Like subject phrases, phrases fulfilling these three functions have several formal properties that make them distinguishable from each other. Compare, for example, the underlined predicatives in (4) and (5).

(4)   He met <u>the President</u>
(5)   He became <u>the President</u>

The predicative *the President* of (4) functions as object of the predicator *met*. By contrast, the predicative *the President* of (5) functions as complement of the predicator *became*. Objects are different from complements in that they can serve as subjects of associated **passive clauses**. Thus there can only be *The President was met by him*, but not **The President was become by him*.

Another important difference is that in contrast to the object, the complement can be expressed by a **bare noun** (i.e. a noun that is used without a

determiner like *the* or *a*) and an adjective. That is, we can say *He became President* and *He became important*, but not **He met President* and **He met important*.

The adjunct is different from both the object and the complement in that it is fairly independent of the predicator with which it combines in the predicate phrase. The adjunct usually provides additional information modifying the action denoted by the predicator – *The French are fighting in Afghanistan* – and is therefore generally non-obligatory. For example, *The French are fighting* is a perfectly grammatical sentence, even though it lacks the place adjunct *in Afghanistan*. By contrast, sentences like **He met* and **He became* seem to be incomplete without an object / a complement like *the President*.

The most important element of a phrase is its **head**; all other elements are its **dependents** (which **modify** or **complement** the meaning of the head). The head of a phrase determines its morphosyntactic properties. For example, the phrase *the French* of (3) is a noun phrase (henceforth NP) because it is headed by the unexpressed but understood noun *soldiers*: *The French are fighting in Afghanistan* means 'The French soldiers are fighting in Afghanistan'. If this were not the case, there would be no justification for the use of the plural *are*. The predicate phrase *are fighting in Afghanistan* is a verb phrase (henceforth VP) because it is headed by the verb *are fighting*. If this were not the case, this phrase would not be able to function as predicate of (3). Finally, the adjunct phrase *in Afghanistan* is a preposition phrase (henceforth PP) because it is headed by the preposition *in*.

Having discussed both the formal properties and the internal structure of (at least some) phrases, let us now proceed to another key concept in syntax, the concept of a sentence. A sentence can be defined as a syntactic unit larger than a phrase. The minimum requirement that a combination of two phrases must fulfill in order to qualify as a sentence is that one of these phrases functions as subject and the other as predicate. A combination of two phrases functioning as subject and predicate forms a **clause**. Sentences consisting of no more than one clause are traditionally called **simple sentences**. For example, the sentence *The French are fighting in Afghanistan* is a simple sentence because it consists of only one clause: as pointed out above, it can only be segmented into the NP *the French* functioning as subject and the VP *are fighting in Afghanistan* functioning as predicate.

Simple sentences must be distinguished from both **complex** and **compound sentences**. Compare, for instance, the sentences (6), (7), and (8).

(6) *This is a good idea* (COCA)
(7) *I don't know if this is a good idea* (COCA)
(8) *[t]his is a good idea and he needs to do it as soon as possible* (COCA)

Like *The French are fighting in Afghanistan*, (6) is an example of a simple sentence. It consists of only one clause *This is a good idea*, in which the NP *this* functions as subject and the VP *is a good idea* as predicate. Sentence (7) is an example of a complex sentence. Here the clause *this is a good idea* is embedded in a larger clause *I don't know if this is a good idea*, in which it functions as object of the VP *don't know if this is a good idea* (Cf. *If this is a good idea is not known by me*). Finally, (8) is an example of a compound sentence. It represents a coordination of two syntactically independent clauses *this is a good idea* and *he needs to do it as soon as possible*.

As will be shown in Chapter 6, some grammatical categories in English are purely syntactic categories. For example, there is only a syntactic but not a semantic difference between the active sentence *He met the President* and the associated passive clause *The President was met by him*: both the former and the latter can be used to refer to one and the same meeting event. Similarly, there is mainly a syntactic difference between the nominative pronoun *he* and the accusative pronoun *him*. Both can refer to one and the same male person, but while the nominative *he* usually fills the subject position (e.g. *He met the President*, not **Him met the President*), the accusative *him* can usually be found in the object position (e.g. *The President met him*, not **The President met he*).

1.2.5 Sociolinguistics

Sociolinguistics is the branch of linguistics which is concerned with free variation at all linguistic levels (i.e. not only at the level of sound, which we discussed in 1.2.3). Free variation can be defined as the semantically irrelevant occurrence of at least two different linguistic units (e.g. sounds, words) in exactly the same environment. For example, as was pointed out in 1.2.3, the word *metal* is pronounced with [t] in Received Pronunciation and with [d] in General American. In a similar way, the word *economic* can be pronounced as either /ˌɛkəˈnɒmɪk/ or /ˌiːkəˈnɒmɪk/.

Following Trubetzkoy (1969: 46), we will classify free variation into **general** and **pathological free variation**. With regard to the latter, consider (9):

(9) *My brother in his room* (from Kaplan 1995: 13)

It is not difficult to see that this clause lacks the auxiliary *be*, connecting the subject NP *my brother* with the predicative PP *in his room*: *My brother in his room* = *my brother is / was in his room*. This is a well-known feature of **African American English**, i.e. one of the non-standard varieties of American English. As Kaplan (1995: 20) points out, the variant *My brother is in his room*, which contains the auxiliary *be*, is not intrinsically superior to *My brother in his room*.

Thus a number of languages (e.g. Standard Russian) do not use auxiliaries in clauses like (9). However, given that in the English-speaking context people who would say *My brother is in his room* are politically more powerful than people who would say *My brother in his room*, the latter variant is regarded as a deviation from the grammatical norm of the English language.

With regard to general free variation, consider (10).

(10)   *It was me who had changed* (COCA)

In this sentence the complement position of the clause *It was me* is filled by the accusative pronoun *me*. In contrast to the subject, which is usually not expressed by the accusative forms *me*, *him*, *her*, *us*, *them*, the complement is often expressed by both the nominative *I*, *he*, *she*, *we*, *they* and the accusative *me*, *him*, *her*, *us*, *them*. Consider, for example, (11).

(11)   *It was I who requested the treatment* (COCA)

In contrast to the non-standard *My brother in his room*, which can be regarded as a 'pathological' deviation from the standard *My brother is in his room*, the use of the accusative *me* in (10) does not represent a 'pathological' deviation from the grammatical norm of the English language: *It was me* of (10) is not less standard than *It was I* of (11). Indeed, as has been established by Biber et al.,

> despite a traditional prescription based on the rules of Latin grammar [i.e. that only a nominative pronoun must be used in a clause like *It was me who...* of (10)], accusative forms are predominant in all registers where the relevant forms are found. (Biber et al. 1999: 335)

Accordingly, the variation between *It was me who...* of (10) and *It was I who...* of (11) is clearly an instance of general free variation.

In addition to the general–pathological distinction discussed above, all instances of free variation can be classified into **stylistically relevant** and **stylistically irrelevant variation** (Trubetzkoy 1969: 47-48). As an illustration of the former, let us again compare the use of the nominative and the accusative forms in sentences like (10) and (11). As just said, Sentence (10), containing the accusative *me*, is not less standard than Sentence (11), containing the nominative *I*. The only difference is that (11) is, from a stylistic point of view, more formal than (10) (Payne and Huddleston 2002: 459). The characteristic 'more formal' always applies to pronouns in the nominative case when they fill the complement position in clauses like *It is I who...* of (11). That is, *It is he who...* is more formal than *It is him who...*; *It is she who...* is more formal than *It is her who...*; *It is we who...* is more formal than *It is us who...*; etc.

By contrast, there does not seem to be a noticeable stylistic difference between the two possible pronunciations of *economic*: neither /ˌekəˈnɑmɪk/ nor /ˌiːkəˈnɒmɪk/ is more / less formal than the other. Similarly, there is no stylistic difference between /ˈmetl̩/ and /ˈmedl/. From the point of view of Received Pronunciation, the American variant /ˈmedl/ can perhaps be regarded as a 'deviation' from the norm. However, it is neither more nor less formal than the standard British pronunciation /ˈmetl̩/. The variation between /ˌekəˈnɑmɪk/ and /ˌiːkəˈnɒmɪk/ as well as between /ˈmetl̩/ and /ˈmedl/ can thus be regarded as instances of stylistically irrelevant free variation.

The categories of stylistically relevant and irrelevant free variation will be of particular importance in Chapters 2 and 3.

1.3 What is a word?

Since morphology defines itself as the study of the internal structure of words, any introduction to morphology must begin with the specification of the **wordhood criteria**, i.e. formal characteristics of words distinguishing them from other linguistic units. To begin with, let us consider the definition of a word provided by Merriam-Webster Online. A word is

> a speech sound or series of speech sounds that symbolizes and communicates a meaning usually without being divisible into smaller units capable of independent use. (MWO)

As we can see, this definition names two wordhood criteria: the **meaning criterion** and the **isolatability criterion**. The former is fairly straightforward: any combination of sounds that is associated with some meaning qualifies as a word. For example, the combination of the sounds /kæt/ is a word because it is associated with the meaning 'cat'; the combination of the sounds /dɪˈmɒkrəsɪ/ is a word because it is associated with the meaning 'democracy'; the combination of the sounds /ˈweɪthʊd/ is a word because it is associated with the meaning 'waithood'; etc.

Unfortunately, not all words (or to be more precise, combinations of sounds which we usually regard as words) have meanings. Consider, for example, the underlined forms in (12), (13), and (14).

(12)   *Do you think he lived a happy life?* (COCA)
(13)   *He took a decision without asking anybody* (COCA)
(14)   *We waited for him to unfold poems like pieces of origami* (COCA)

Objects like the NP *a happy life* of (12) are traditionally called **cognate objects**: they are headed by nouns which are lexically cognate with the predicator verbs with which they combine in the same predicate VP (e.g. *to live* and *a life*, *to die* and *a death*, *to sleep* and *a sleep*, etc.). From a semantic point of view, cognate objects are different from other object NPs in that their head nouns do not make much contribution to the meanings of the VPs in which they occur. Indeed, *He lived a happy life* means 'he lived happily'; *He died a painful death* means 'he died painfully'; *He slept a restless sleep* means 'he slept restlessly'; etc.

Similarly, the predicator verb *took* of (13) does not make much contribution to the meaning of (13): *He took a decision without asking anybody* means 'he decided without asking anybody'. Verbs like the *took* of (13) are known as **light verbs**. Apart from *take*, light verbs include *make* of *make a decision* 'to decide', *give* of *give a kiss* 'to kiss', *have* of *have a drink* 'to drink', *put* of *put the blame on somebody* 'to blame somebody', etc. All these verbs are 'light' in that they carry only grammatical meanings such as e.g. 'the past tense' of *took* in (13), but do not denote any actions or states, which is characteristic of other verbs as well as of non-light uses of the same verbs. That is, for example, while *gave* of *She gave him a book* is a 'normal' verb carrying the action meaning 'to give, to transfer', *gave* of *She gave him a kiss* is light verb carrying only the grammatical meaning 'the past tense'.

Finally, in contrast to the preposition *for* of *I did it for him*, which carries the meaning 'in defense or support of; in favor of, on the side of' (OED), *for* of (14) does not carry any meaning of its own but only serves to connect the predicator *waited* with the object pronoun *him*. Verb + preposition combinations like *wait for somebody*, *account for something*, *depend on somebody / something*, *testify to something*, etc. are traditionally called **prepositional verbs**. These are different from other verb + preposition combinations with regard to a number of properties. One is the semantic emptiness of an accompanying preposition. Thus there does not seem to be a particular reason why speakers of English *wait for somebody*. In German, we *wait on somebody* (*warten auf jemanden* → wait on somebody). In Russian, there is no preposition at all (*ждать кого-то* / *zhdat kogo-to* → wait somebody). That *for* does not make any contribution to the meaning of *wait for somebody* can also be corroborated by the existence of the verb *await*, which expresses a similar meaning without any preposition; e.g. (15).

(15)   I *awaited* their critique (COCA)

To conclude: if there are words like *life* of (12), *took* of (13), and *for* of (14) that express no (lexical) meanings, then the meaning criterion cannot be a reliable wordhood criterion. (An alternative to this conclusion would be to regard the forms under analysis as non-words. But this analysis seems to go against our feelings: in this case, we have to distinguish between e.g. the word *take* of *He took a book out of his pocket* and the non-word *take* of *He took a decision*.) In addition to this, if we want to regard the word as a formal category, we must try to identify at least one formal property of words making them distinguishable from (instances of) other formal categories.

Now, let us proceed to the isolatability criterion. This criterion defines a word as a **minimum free form**, i.e. a form capable of independent use. According to Bloomfield (1973[1934]: 178), 'independent use' means that the form in question can form a one-word **elliptical sentence**, i.e. a sentence which does not have the usual subject + predicate structure (e.g. *He saw her*) but consists of only one word (e.g. *He.*). Consider, for example, the underlined forms *black* of (16) and *cat* of (17).

(16) *Pamela remains buried in her magazine, and Susan writes, Lauren. Five. Karate-trained daughter. Miniature of her diminutive mother. Fine, blond hair, pasted to nose and mouth. Scowls at turkey sandwich. Could this small girl be a black belt...? Hmm. <u>Black</u>? Susan crosses it out* (COCA)

(17) *Man I'm trying to sleep. <u>Cat</u>? <u>Cat</u>? I'll get the cat* (COCA)

Both *black* and *cat* in the examples above form one-word elliptical sentences and therefore qualify as words in accordance with the isolatability criterion. In (16), the elliptical interrogative *Black?* can be expanded into *Does this small girl really possess a black belt?*, whereas *Cat?* of (17) is expandable into *Where is the cat that is disturbing me at the moment?*

In addition to adjectives and nouns, the isolatability requirement is also fulfilled by pronouns, verbs (including auxiliaries), adverbs, conjunctions, and interjections. Consider, for example, the underlined forms in (18), (19), (20), (21), (22), and (23).

(18) *And you know who that makes the ultimate winner? <u>Us</u>* (COCA)

(19) *Here are things she said he did: Kept secrets carefully. Knew who he was. Believed in God. <u>Gave</u>. And gave. Studied hard. <u>Thought</u>. Took life seriously* (COCA)

(20) *'She was in Mrs. Potter's class,' said Peg Leg. <u>Was</u>? I thought, feeling a stab of panic. I'd had Mrs. Linda Potter last year [...]* (COCA)

(21) *He said, 'Would you like to know something? I went to Woodstock.' She looked as though he'd started speaking Vietnamese. 'No kidding. <u>Recently</u>? The movie The town? What?* (COCA)

(22) VALENTINE: *As you know, her brother died.*
ORSINO: *Yes. <u>And</u>?*
VALENTINE: *She plans to cry for the next seven years* (COCA)

(23)   They want no oil. <u>Wow</u>! That leaves us with wind (COCA)

The elliptical accusative pronoun *Us* of (18) can be easily expanded into *That makes us the ultimate winner*.

The elliptical verbs *Gave* and *Thought* of (19) can be expanded into *He gave* and *He thought*.

The elliptical auxiliary *Was?* of (20) can be expanded into *Is it really the case that she <u>was</u> in Linda Potter's class? I think she is still in this class*.

The elliptical adverb *Recently?* of (21) can be expanded into *Is it really true that you have recently gone to Woodstock?*

The elliptical conjunction *And?* of (22) can be expanded into *And how is her brother's death related to what we are now talking about?*

Finally, the elliptical interjection *Wow!* of (23) can be expanded into *Wow, that's great that they want no oil!*

All these forms fulfill the isolatability criterion and, accordingly, qualify as words.

Now, let us also discuss the status of the articles *a* and *the*. In contrast to the word classes named above, the forms in question do not seem to be capable of independent use in English. As Bloomfield (1973[1934]: 179) points out, "we can imagine a hesitant speaker who says *The...* and is understood by his hearers". However, such sentences rarely, if ever, occur in real life: the largest balanced corpus of contemporary American English – the already mentioned COCA, from which most of the examples used in this book are drawn – does not contain a single elliptical sentence made up of *a* and *the* only. This fact may lead us to the conclusion that the articles in English are non-free forms and hence cannot be regarded as words. In this connection, Bloomfield observes that

> the form *the*, though rarely spoken alone, plays much the same part in our language as the forms *this* and *that*, which freely occur as sentences; this parallelism leads us to class *the* as a word. (Bloomfield 1973[1934]: 179)

In other words, both *the* and *this* of e.g. the NPs *the cat* and *this cat* function as determiners, marking the head noun *cat* as definite. Given this functional similarity between *this* and *the* and given that *this*, in contrast to *the*, is capable of independent use (e.g. (24)) and thus qualifies as a word, we are justified in concluding that *the* is a word as well.

(24)   Stevie was pulling on his shirt. His pale skin had turned bright pink, but before he tugged down the shirt I saw a dark stripe on his chest. What the hell is that? I said. <u>This</u>? He lifted his shirt (COCA)

The same line of reasoning enables us to analogize the indefinite article *a* to the word *some*: like *a* of the NP *a cat*, *some* of the NP *some cat* functions as determiner, marking the head noun *cat* as indefinite. Given this functional similarity between *a* and *some* and given that *some*, in contrast to *a*, is capable of independent use and thus qualifies as a word, we are justified in concluding that *a* is a word as well.

In addition to the meaning and isolatability criteria, many studies also mention:

- **the movement criterion**
- **the uninterruptability criterion**
- **the orthographic criterion**

The movement criterion defines a word as a form that can move within a clause. Unfortunately, in a language like English, which has a **fixed word-order** (recall that the default position of the subject in English is before the main verb), this criterion is of very little help. Consider, for example, (25)-(29).

(25)   *She gave him a kiss*
(26)   **Gave she him a kiss*
(27)   **Gave him she a kiss*
(28)   **Gave him a she kiss*
(29)   **Gave him a kiss she*

The **ungrammaticality** of (26)-(29) illustrates that the subject *she* of (25) cannot move within the sentence and thus does not qualify as a word. Likewise, as (30)-(32) illustrate, the predicator verb *give* also does not fulfill the movement requirement and thus cannot be regarded as a word.

(30)   **She him gave a kiss*
(31)   **She him a gave kiss*
(32)   **She him a kiss gave*

The only possible transformations of (25) are (33) and (34).

(33)   *A kiss, she gave him*
(34)   *Him, she gave a kiss*

Accordingly, we must conclude that (25) consists of the two non-words *she* and *gave* and the two words *him* and *a kiss*. Needless to say, this is an extremely counter-intuitive conclusion.

The uninterruptability criterion defines a word as a string of sounds that does not allow the insertion of a modifying element within its boundaries. This criterion was proposed by Bloomfield (1973[1934]: 180) for distinguishing between non-idiomatic adjective + noun NPs like *a black board* and idiomatic adjective + noun complex words like *blackboard*. Of the two, only the former allow the insertion of a modifier. That is, for example, we can say *a black – that is, bluish-black – board* but not **a blackthatisbluishblackboard*.

Like the movement criterion, the uninterruptability criterion has a serious drawback: it provides no justification for distinguishing between words and some fully-idiomatic VPs. Consider, for example, the VP *shoot the breeze* 'to chat, talk idly' (OED). Just like *blackboard*, *shoot the breeze* is uninterruptable in that the component *breeze* cannot be modified by adjectives like *cool, warm, gentle, soft, fresh, sudden*, etc. which modify *breeze* in non-idiomatic contexts (e.g. *I felt the warm breeze in my face*). We cannot say **shoot the cool breeze*, **shoot the warm breeze*, **shoot the sudden breeze*, etc. Accordingly, the VP *shoot the breeze* fulfills the uninterruptability criterion and thus must be regarded as a word.

Finally, the orthographic criterion defines a word as a string of letters that is separated from other words by means of a blank space. For example, the non-idiomatic NP *a black cat* can be segmented into three words – *a*, *black*, and *cat* – simply because in this NP there are blank spaces separating *a* from *black* and *black* from *cat*. (It does not really matter that of these components, only *black* and *cat* fulfill the isolatability criterion discussed above.) The same applies to the idiomatic VP *shoot the breeze*. The VP under consideration can be said to consist of three words – *shoot*, *the*, and *breeze* – simply because we do not write **shootthebreeze*. Finally, recall the semantically empty preposition *for* of *wait for somebody*. Again, we can argue that *for* qualifies as a word simply because we write *wait for somebody* rather than **waitforsomebody*.

The presence of a blank space seems to be the easiest way of determining whether a particular string of letters represents a word or not. But it cannot be regarded as a sufficient wordhood criterion. The major problem with the orthographic criterion is not the frequently mentioned instability of the spelling of some English words – an often cited example is *flowerpot*, which can be spelled *flowerpot, flower-pot*, and *flower pot* – but the secondary and the artificial character of the writing system in relation to spoken speech (see e.g. Lieber and Štekauer 2009a: 7). The former appeared much later than the latter; e.g. the orthographic forms *flowerpot / flower-pot / flower pot* appeared in the English language later than the sound form /ˈflaʊəpɒt/. Hence it is extremely doubtful that word boundaries can be determined only on the basis of orthographic conventions. This position has been recently advocated by Haspelmath (2011), who argues that

[m]any orthographies, especially (but not only) those based on the Greek, Latin and Cyrillic alphabets, use spaces between words. However, there are also many orthographies that do not use spaces, e.g. Chinese, Japanese, and Sanskrit. [...] In the European languages, too, word spacing is an innovation; until about a thousand years ago, *scriptio continua* (continuous writing) was the norm in Western writing [...]. There is no doubt that the modern orthographic use of spaces is to some extent guided by language structure, but not in such a way that conventional spelling could be used to decide contentious issues. (Haspelmath 2011: 36)

To conclude: of the five criteria discussed above, only the isolatability criterion provides a relatively unproblematic formal basis for distinguishing between words and instances of other formal categories (i.e. forms that are not capable of independent use). In accordance with this criterion, a word is any combination of sounds which can form a one-word elliptical sentence such as *Black?* of (16), *Cat?* of (17), *Us* of (18), *Gave* and *Thought* of (19), *Was?* of (20), *Recently?* of (21), *And?* of (22), and *Wow!* of (23). (The isolatability criterion does not provide a basis for distinguishing between NPs like *a black board* and complex words like *blackboard*: both the former and the latter can be analyzed as combinations of two words, which are capable of independent use. We will return to this issue in Section 5.6.)

The conclusion drawn above leads us to a very important practical question: how can we actually establish whether some particular combination of sounds is capable of occurring as a one-word sentence? There are two answers to this question. If you are a native speaker of English, you can try to invent a context (similar to those of the examples above) in which the form in question occurs as an elliptical sentence. If you succeed in that, you will be justified in claiming that the form under analysis does indeed qualify as a word. For example, if you, like Bloomfield, can imagine a hesitant speaker who says *The...* or *A...* and is understood by the hearers, you can claim that the articles in English are words.

Alternatively, this question can be answered by looking at actual language use, i.e. consulting a balanced linguistic corpus like Corpus of Contemporary American English / COCA (http://www.americancorpus.org/) or British National Corpus / BYU-BNC (http://corpus.byu.edu/bnc/). Just enter e.g. . take . (with spaces separating *take* from the full stops) to the search mask of the corpus of your choice and click at 'Search'. Both COCA and BYU-BNC will then search for the occurrences of *take* in which it is preceded and followed by a full stop. One of such occurrences could be an elliptical sentence like .*Gave.* of (19). If you enter . *take* ? or . *take* ! (also with spaces separating *take* from the punctuation marks), the corpora will search for the occurrences of *take* in which it is preceded by a full stop and followed by either a question mark or an exclamation mark. One of such occurrences could be a one-word interrogative sentence like .*Black?* of (16) or a one-word exclamatory sentence like .*Wow!* of (23). If you succeed in finding at least one elliptical sentence containing *take*, you will be justified in regarding *take* as a word.

If neither COCA nor BYU-BNC yield elliptical sentences consisting of the form in question, this does not necessarily mean that it cannot be regarded as a word. It may be the case that the databases of both corpora simply do not contain elliptical sentences consisting of the form in question. For instance, neither COCA nor BYU-BNC contain elliptical sentences consisting of the form *baboon*. This fact may lead us to the conclusion that *baboon* does not fulfill the isolatability criterion and, accordingly, does not qualify as a word. However, at an earlier point, we established that the isolatability criterion is fulfilled by a semantically related word *cat*: *Man I'm trying to sleep. Cat? Cat? I'll get the cat* of (17). We can imagine a similar context involving a person who is trying to fall asleep and a baboon which is somehow disturbing that person. The former gets angry and says *Man I'm trying to sleep. Baboon? Baboon? I'll get the baboon*. This is sufficient to regard *baboon* as a word.

1.4 Exercises

1. Make sure you can explain each of the key terms printed in boldface (ideally, using your own examples).

2. Which of the following statements are true?

a) A complex word consists of only one formally indivisible component.
b) A concept is a mental description that is activated by a particular sound form.
c) Opacity is one of the characteristics of literal meaning.
d) Acoustic phonetics is concerned with the production of sounds.
e) Sounds that realize two different phonemes are in contrastive distribution.
f) Phrases never perform syntactic functions.
g) Complement phrases can function as subjects of associated passive clauses.
h) Free variation can either be stylistically relevant or stylistically irrelevant.
i) Light verbs express no lexical meanings.
j) A word is a linguistic unit which is not capable of independent use.

3. State which of the following words are complex words. Explain your analysis.

a) *dog*
b) *untrue*

c) *freedom*
d) *stage*
e) *democracy*
f) *to disambiguate*
g) *to satisfy*
h) *mortgage*
i) *defendant*
j) *chair*

4. Name the syntactic function which the underlined phrases perform in the sentences below. Explain your analysis.

a) *He <u>broke the record</u>* (COCA)
b) *<u>It</u> is hot for September* (COCA)
c) *I kept <u>quiet</u>* (COCA)
d) *<u>Yesterday</u> I turned twenty-nine* (COCA)
e) *He stood <u>in the doorway</u>* (COCA)
f) *So I wrote <u>a letter</u>* (COCA)
g) *That data <u>became even more important</u>* (COCA)
h) *<u>To err</u> is human* (COCA)
i) *I helped <u>him</u>* (COCA)
j) *<u>In case of emergency</u>, you must do the following* (COCA)

5. Using COCA, find out which of the following forms qualify as words in accordance with the isolatability criterion.

a) *she*
b) *truly*
c) *girls*
d) *has*
e) *to*
f) *stand*
g) *under*
h) *three*
i) *where*
j) *damn*

1.5 Further reading

For an overview of how different linguists define the term 'morphology', see Heringer (2009: 9-18).

Löbner (2002) and Cruse (2004) are good introductory textbooks to semantics: both contain very detailed discussions of such important concepts as 'meaning', 'conceptual category', 'semantic decomposition', 'necessary and sufficient conditions', 'prototype', etc.

A recent introduction to English phraseology is Fiedler (2007). For an overview of the contemporary research in phraseology, see Burger et al.'s (2007) *Handbook of Phraseology* as well as the recent monographs by Dobrovol'skij and Piirainen (2009; 2005).

There are many good introductions to English phonetics and phonology. See, for example, Gut (2009), Cruttenden (2008), Yavaş (2006), McMahon (2004). Trubetzkoy's (1939) *Grundzüge der Phonologie* is a classic monograph dealing with such key issues as differences between phonetics and phonology, identification of phonemes, distribution of sounds, etc. The English translation is Trubetzkoy (1969).

Huddleston and Pullum (2005) is an excellent introduction to English syntax discussing such issues as syntactic functions and their formal properties (e.g. subject-hood criteria, differences between complements, objects, and adjuncts), the internal structure of phrases in English, types of sentences in English, etc. A more detailed discussion of the same issues can be found in their earlier work *The Cambridge Grammar of the English Language* (Huddleston and Pullum 2002). A more recent introduction to English syntax is Kreyer (2010).

Many authors (e.g. Aronoff and Fudeman 2005: 33-44) introduce a number of subcategories of words: for example, **orthographic words, phonological words, morphosyntactic words**, etc. This approach is typical of authors who conclude that words cannot be distinguished from non-words on the basis of one criterion such as, for example, the isolatability criterion. Since this book takes a different approach (namely, that most forms that language users regard as words can be identified as words on the basis of the isolatability criterion), it will not use any of these terms.

2 The internal structure of English words

This chapter focuses on the morphemic segmentation of English words as well as on the typology and distribution of morphs in English. Section 2.1 provides a very basic definition of a morpheme as the smallest meaning-carrying unit. Section 2.2 analogizes the concept of a morpheme to that of a linguistic sign and, on the basis of this, formulates a more precise definition of a morpheme as a conventionalized association between a particular form and a particular meaning. Section 2.3 is concerned with the distribution of morphs in English: the main question here is when two different forms can be regarded as allomorphic realizations of the same morpheme rather than as forms realizing two different morphemes. Section 2.4 deals with the principles of morphemic segmentation: how can we decide whether the word under analysis can be segmented into (at least) two morphemes or is a monomorphemic word non-segmentable into smaller meaning-carrying units? Finally, Section 2.5 discusses the hierarchy of different morpheme types. In this section, we will learn, among other things, in which respects roots are different from affixes and whether so-called combining forms constitute a distinct morpheme type in English.

2.1 What is a morpheme?

A morpheme is traditionally defined as the smallest meaningful component of a word. As an illustration of this definition, let us consider the meaning of *untrue* in (35).

(35)   Her lover had been <u>untrue</u> (MWO)

According to MWO, the *untrue* of (35) means 'not faithful'. It is evident that the meaning of *untrue* represents the reverse of the meaning of *true*. Thus the sentence *Her lover had been true* means the opposite of (35), namely, that the subject *her lover* was faithful to his lover in that he did not have sexual relations with a third person. Hence the word *untrue* can be segmented into two meaning-carrying units: {un} and {true}[3]. The unit {true} carries the meaning 'faithful'; the unit {un} carries the meaning 'negation of the following adjective *true*'. In a similar way, the word *unable* can be segmented into 1) the unit {able} carrying the meaning 'able', i.e. 'having sufficient power, skill, or resources to accomplish

---
[3] Linguists usually put morphemes in curly braces {}.

an object' (MWO) and 2) the unit {un} carrying the meaning 'negation of the following adjective *able*'.

Units like {un} and {true} of *untrue* and {un} and {able} of *unable* are morphemes. They are the smallest meaningful components of the words under analysis, i.e. components which cannot be further segmented into smaller meaning-carrying units. That is, for example, the unit *un-* cannot be further segmented into the morphemes {u} and {n} because the sounds [ʌ] and [n] do not carry any discernible meanings of their own. Similarly, *true* cannot be further segmented into the morphemes {tr} and {u} because the sounds [tr] und [u:] do not mean anything.

In Section 1.1 we became acquainted with the distinction between simple and complex words. Given what we have just learned about morphemes, we are now in a position to give a more precise definition of the terms 'simple word' and 'complex word'. Simple words are **monomorphemic words**, i.e. words like *cat* that can be segmented only into one morpheme. Complex words, by contrast, are **polymorphemic words**, i.e. words like *untrue*, *unable*, and *waithood* which can be segmented into at least two morphemes.

Morphemes are traditionally classified into a number of categories. With regard to their autonomy, morphemes are usually classified into **free** and **bound morphemes**. Free morphemes fulfill the isolatability requirement which we discussed in 1.3. For example, the simple word *cat* can be said to consist of the free morpheme {cat} because, as we established in 1.3, the sound form /kæt/ can form a one-word elliptical sentence. The same applies to the morpheme {true} of *untrue*. By contrast, the morpheme {un} of *untrue* is a bound morpheme, i.e. a morpheme which does not fulfill the isolatability requirement: we cannot form an elliptical sentence consisting of *un-* only.

With regard to their function, morphemes are classified into **lexical** and **grammatical morphemes**. Lexical morphemes are morphemes like {cat} of *cat* and {un} of *untrue* which express **optional lexical meanings**, i.e. meanings which are expressed only when language users specifically want them to be expressed. Grammatical morphemes are morphemes like {s} of *books* and {ed} of *I worked* which express **obligatory grammatical meanings**, i.e. meanings which cannot be unexpressed. For example, in English nouns are always marked with regard to the grammatical category NUMBER (i.e. nouns are either in the singular or in the plural number) and verbs are always marked with regard to the grammatical category TENSE (i.e. they are either in the present or in the past tense).

With regard to their form, morphemes can be classified into **continuous** and **discontinuous morphemes**. The latter are morphemes which are interrupted by other morphemes. Consider, for example, the grammatical meaning 'the progressive aspect' of *I am working*. It is evident that this meaning is inherent in both the free auxiliary *be* and the bound form *-ing*. If we remove these forms, the clause under analysis will acquire a different grammatical meaning: *I work* is in the non-progressive aspect. Accordingly, we can conclude that the progressive meaning of *I am working* is expressed by the discontinuous morpheme {am...ing}, which is interrupted by the lexical morpheme {work}. The majority of morphemes in English and other languages are, however, continuous morphemes, i.e. morphemes like {cat} of *cat*, {un} and {true} of *untrue*, {wait} and {hood} of *waithood*, etc. which are made up of "sequences of consecutive phonemes" (Harris 1945: 121).

Usually, morphemes carry only one meaning. That is, for instance, the morpheme {cat} of *cat* carries the meaning 'cat'; the morphemes {un} and {true} of *untrue* carry the meanings 'not' and 'faithful'; the morphemes {wait} and {hood} of *waithood* carry the meanings 'to wait' and 'stage'; etc. Some morphemes, however, simultaneously express more than one meaning. Consider, for instance, the meanings inherent in the bound morpheme *-s* of *runs* in e.g. *He runs*. It is evident that this morpheme carries several grammatical meanings. These include:

- 'the present tense' (cf. *He runs* and *He ran*)
- 'the non-progressive aspect' (cf. *He runs* and *He is running*)
- 'the active voice' (cf. *He runs* and *This business was run by him*)
- 'the indicative mood' (cf. *He runs* and *Run!*)
- 'the third person' (cf. *He runs* and *You run*)
- 'the singular number' (cf. *He runs* and *They run*)

Morphemes like *-s* of *runs* which simultaneously express more than one meaning are often called **portmanteau morphemes** (or **mega-morphemes** or **cumulative morphemes**).

We will enlarge on the typology of morphemes in Sections 2.4 and 2.5.

2.2 Morphemes as signs

A morpheme is also very often defined as the smallest **linguistic sign**, i.e. a conventionalized association between a particular form and a particular meaning (also commonly referred to as the **signifier** and the **signified**). Analyzing the previous example {true}, we can say that in the English language there exists a conventionalized association between the sound form /tru:/ and the concept of being true. That is, when speakers of English hear the word *true* in sentences like (35), they think of a person who is faithful to another person (in a sexual sense). The morpheme {true} thus represents a linguistic sign made up of the signifier /tru:/ and the signified 'faithful', there being a conventionalized association between the former and the latter.

2.2.1 One signifier → more than one signified

Very often, however, one and the same signifier is associated with more than one signified. In this case, we are dealing with either **polysemy** or **homonymy**. The hallmark of polysemy is that two or more signifieds inherent in the same signifier have something in common. A good example is the *true* of (36).

(36)   Indicate whether each of the following statements is <u>true</u> or false. (MWO)

In contrast to the *true* of (35), the *true* of (36) does not mean 'faithful' but 'not false' (MWO). The fact that one and the same signifier *true* can express both the meanings 'faithful' and 'not false' is clearly an instance of polysemy because the meaning 'not false' can be easily derived from the meaning 'faithful': if a statement is not false, its speaker can be said to be faithful to the truth. The sense 'not false' can thus be seen as a product of **semantic narrowing** of the sense 'faithful': 'faithful' > 'faithful to the truth, not false'.

Homonymy is different from polysemy in that two or more signifieds which can be expressed by the same signifier do not have much in common. Compare, for example, the meanings of *case* in (37) and (38).

(37)   In this <u>case</u>, there's no question (COCA)
(38)   Lifting a heavy <u>case</u>, Nick places it into the cab's trunk (COCA)

In contrast to the meanings 'faithful' and 'not false' of *true*, the two meanings of *case* exemplified by (37) and (38) – 'a set of circumstances or conditions' (MWO) and 'a box or receptacle for holding something' (MWO) – do not seem to have anything in common. Accordingly, their co-existence in the signifier *case* must be regarded as an instance of homonymy, not polysemy.

2.2.2 One signified ← more than one signifier

The reverse situation is possible as well: one and the same signified can be expressed by more than one signifier. For example, the concept of a person who has sex only with his or her regular sexual partner (i.e. spouse, girlfriend) can be expressed by both *true* and *faithful*. Cf. (39) and (40).

(39)   Her boyfriend had been <u>true</u>
(40)   Her boyfriend had been <u>faithful</u>

Accordingly, the signifier *faithful* can be considered a **synonym** of the signifier *true*.

A somewhat similar example is the word *detail*, which can be pronounced (by one and the same speaker of General American) either /ˈdiːteɪl/ or /dɪˈteɪl/. That is, the first syllable can either be stressed and end in the long vowel [iː] or be unstressed and end in the short vowel [ɪ]. Similarly, *doctrinal* can be pronounced both /ˈdɒktrɪnəl/ and /dɒkˈtraɪnəl/. That is, the second syllable can either be unstressed and end in the short vowel [ɪ] or be stressed and end in the diphthong [aɪ]. (According to MWO, the pronunciation /dɒkˈtraɪnəl/ is, however, more typical of Received Pronunciation than of General American.)

As we have learned in the previous chapter, in phonetics and phonology cases like *detail* and *doctrinal* are regarded as instances of free variation (or free distribution) of sounds. Free variation is similar to synonymy in that in both cases the same meaning is expressed by more than one sound form. However, in the case of free variation, there remains a considerable phonetic similarity between the forms in question: as pointed out above, /ˈdiːteɪl/ and /dɪˈteɪl/ differ only with regard to the placement of stress and the length of the first vowel. Similarly, /ˈdɒktrɪnəl/ and /dɒkˈtraɪnəl/ differ with regard to the placement of stress and the quality of the second vowel. Conversely, in the case of synonymous forms like *true* and *faithful*, there are no or very few phonetic similarities. Compare, for example, /truː/ and /ˈfeɪθfʊl/. These words do not share a single sound.

2.2.3 The syntactics of a sign

The conception of a linguistic sign as a conventionalized association between a particular signifier and a particular signified was formulated by the Swiss linguist Ferdinand de Saussure (1857-1913) in his posthumously published *Cours de Linguistique Générale / Course in General Linguistics* (1973: 97-103). An important revision of this conception was made by the Russian-Canadian linguist Igor Mel'čuk (1979: 171; cf. 1968: 425), who, inspired by the works of the American semiotician Charles Morris, defined the linguistic sign as a triplet consisting of the following three elements:

1. **the signifier**
2. **the signified**
3. **the syntactics**

The latter component – the syntactics of the sign – is described by Mel'čuk (1979: 171) as "the set of all data about the combinatorial properties" including "part of speech, [...] declension or conjugation type, phonological and / or morphological environments, selectional restrictions of all kinds, etc.". Compare, for example, the adjective *happy* and the adverb *happily*. As regards their signifieds, it may seem that the forms under analysis express different meanings: while *happy* means 'happy', *happily* means 'in a happy manner'. But compare the meanings of the clauses (41) and (42), containing the adjective *happy* and the adverb *happily*.

(41)   *He lived a happy life*
(42)   *He lived happily*

As has been recognized by many authors (e.g. Pullum and Huddleston 2002: 529), adjectives like *happy* and morphologically related adverbs like *happily* express essentially the same meaning. Indeed, both *He lived a happy life* and *He lived happily* can be paraphrased by *His life was happy*. Accordingly, the difference between *happy* and *happily* cannot be described in terms of their signifieds. What distinguishes the former from the latter is that the adverb *happily* cannot fill two syntactic positions that are typically filled by the corresponding adjective *happy*. First of all, *happily* cannot function as modifier of a head noun in an NP like *a happy life*. That is, we cannot say **a happily life*. Second, *happily* cannot function as complement in a sentence like *His life was happy*: there can only be *His life was happy* but not **His life was happily*.

Obviously, a linguistic theory that defines a sign as a doublet consisting of the signifier and the signified cannot explain the difference between an adjective like *happy* and an adverb like *happily*. As, for example, Pullum and Huddleston (2002: 529) argue, "it is function that provides the primary basis for the distinction between adjectives and adverbs". But what precisely is *function*? Is function also a component of a linguistic sign? By contrast, a theory that defines the sign as a triplet consisting not only of the signifier and the signified but also of the syntactics can easily explain the difference: the adjective *happy* and the adverb *happily* differ from each other with regard to their syntactics.

2.2.4 The sociolinguistics of a sign

There is no doubt that the inclusion of the syntactics represents a significant improvement of Saussurean conception of a linguistic sign. However, as we will see below, signs also have properties that cannot be attributed to either their syntactics or their signifieds.

Compare, for example, the forms *fever* and *pyrexia*. Both of them express the same meaning: 'abnormal elevation of body temperature' (MWO). Both of them can head NPs that function as subjects or predicatives in predicate VPs. E.g. (43) and (44).

(43)   Jaden had *a fever* (COCA)
(44)   A wound swab was taken as Julie still had *a pyrexia* (BYU-BNC)

Accordingly, these forms do not differ with regard to either their signifieds or their syntactics. The difference between *fever* and *pyrexia* is that of **register**: while the former is a stylistically neutral term that can be used in all kinds of contexts, the latter is a highly professional term that occurs in highly specialized contexts: COCA has only 6 occurrences of *pyrexia* in scientific articles published in medical journals and 6276 occurrences of *fever* in all possible contexts. The free variation between *fever* and *pyrexia* is thus an instance of stylistically relevant free variation.

Recall also that the word *doctrinal* can be pronounced both /ˈdɒktrɪnəl/ and /dɒkˈtraɪnəl/, the latter variant being more typical of Received Pronunciation than of General American. Again, as in the case of *fever* and *pyrexia*, we observe a sociolinguistic (not a syntactic or a semantic) variation between the forms in question. That is, /ˈdɒktrɪnəl/ and /dɒkˈtraɪnəl/ do not differ with regard to either their signifieds or their syntactics: both signifiers express the same meaning 'doctrinal' and can fill the syntactic positions that are typically filled by adjectives. The only difference is that the variant /dɒkˈtraɪnəl/ is more likely to be used by a speaker of Received Pronunciation, whereas a speaker of General American will most likely give preference to the variant /ˈdɒktrɪnəl/.

Taking all this into account, we can conclude that a linguistic sign represents a conventionalized association between the following four elements:

1. **the signifier**
2. **the signified**
3. **the syntactics**
4. **the sociolinguistics**

The sociolinguistics of a sign is the component that contains the information as to whether the signifier of the sign

- is stylistically neutral, formal, colloquial, slang, etc.
- is mainly used by people of a particular profession, social status, gender, etc.
- is characteristic of a particular variety of English such as American English, British English, Australian English, etc.

2.2.5 The signified as the most important sign component

The definition of a linguistic sign given above raises the question of which of the four components of the sign – the signifier, the signified, the syntactics, the sociolinguistics – must be seen as its most important component. The answer to this question has important implications for our analysis of the distribution of signifiers like *happy* and *happily* that have different syntactics but identical signifieds: if we decide that the syntactics of a sign is more important than its signified, we will be justified in claiming that the signifiers *happy* and *happily* are in contrastive distribution and, accordingly, can be regarded as instances of two different signs. Similarly, if we decide that the sociolinguistics of a sign is more important than its signified, we will be justified in claiming that *fever* and *pyrexia* occur in contrastive distribution and thus realize two different signs. By contrast, if we decide that the signified is more important than both the syntactics and the sociolinguistics, we will have to analyze *happy* and *happily* as instances of the same sign that occur in complementary distribution[4] and *fever* and *pyrexia* as instances of the same sign that occur in stylistically relevant free variation.

This book argues for the latter solution. It is the signified (not the syntactics or the sociolinguistics) of a sign that must be seen as its most important component. This is so because the semantic function is a much more important motivation for the use of linguistic signs than both the syntactic and the sociolinguistic function. That is, the primary reason why we use the signifiers *happily*, *pyrexia*, /dɒkˈtraɪnəl/, etc. is not our wish to either fill particular syntactic positions which the syntactics of these signifiers allow them to fill or to mark these signifiers as stylistically neutral, formal, characteristic of a particular variety of English, etc. We use these signifiers because, first and foremost, we want to express the meanings that are conventionally associated with these signifiers in English.

Accordingly, given the primacy of the signified over both the syntactics and the sociolinguistics (and the signifier), we can conclude this section with the following claims:

- Two different signifiers that are conventionally associated with two different signifieds – e.g. the signifiers *detail* and *doctrinal*, which are associated with the meanings 'detail' and 'doctrinal' – are instances of two different signs. They are in contrastive distribution.

- Two identical signifiers that are conventionally associated with two different signifieds – e.g. the signifier *true*, which is associated with the meanings 'faithful' and 'not false' – are likewise instances of two different signs. They are also in contrastive distribution.

---

[4] Recall that there can only be the NP *a happy life* and the clause *His life was happy*, but not **a happily life* and **His life was happily*.

- Two different signifiers that are conventionally associated with one and the same signified are instances of the same sign, regardless of the differences between these signifiers concerning either their syntactics or their sociolinguistics. Signifiers like *happy* and *happily* which differ from each other with regard to their syntactics are in complementary distribution. Signifiers like *fever* and *pyrexia* which differ from each other with regard to their sociolinguistics are in free variation.

2.3 The distribution of morphs

In morphology the signifier of a morpheme is often referred to as its **morph**. Similar to phones that realize different phonemes, morphs that realize different morphemes (i.e. those that have different meanings) can be said to occur in contrastive distribution. For example, the morphs /truː/ and /ˈeɪ.bl/ of the morphemes {true} and {able} are in contrastive distribution because they carry two different meanings: 'true' and 'able'.

Like in phonetics and phonology, contrastive distribution of morphs that realize two different morphemes must be distinguished from both free and complementary distribution of morphs that realize the same morpheme. As pointed out in the previous section, one and the same signified can sometimes be expressed by more than one signifier. Consider again the two possible pronunciations of *detail*: /ˈdiːteɪl/ and /dɪˈteɪl/. Given that one and the same speaker of American English can pronounce *detail* on one occasion as /ˈdiːteɪl/ and on another occasion as /dɪˈteɪl/ and given that the word *detail* cannot be segmented into smaller meaningful units[5], we are justified in concluding that the morpheme {detail} can be realized by the morphs /ˈdiːteɪl/ and /dɪˈteɪl/ occurring in free distribution. Morphs that realize one and the same morpheme are called **allomorphs**. E.g. the morphs /ˈdiːteɪl/ and /dɪˈteɪl/ are the two allomorphs of the morpheme {detail}.

As regards complementary distribution, consider the words *inadequate* and *impossible*. Since the meanings of these words are the opposites of the meanings of *adequate* and *possible*, we can conclude that the negative meaning 'not' is inherent in the morphs /ɪn/ and /ɪm/. What is important here is that unlike /ˈdiːteɪl/ and /dɪˈteɪl/ of *detail*, /ɪn/ and /ɪm/ are not in free variation: we can say neither */ɪmˈædɪkwət/ instead of /ɪnˈædɪkwət/ nor */ɪnˈpɒsɪb(ə)l/ instead of /ɪmˈpɒsɪb(ə)l/. Similar to complementary distribution of allophones of the same phoneme, complementary distribution of allomorphs of the same morpheme involves a morpheme that is realized by at least two allomorphs which occur in different environments, thereby complementing each other. In the case of the

---
[5] In contrast to the morph *de-* of e.g. *destabilize*, the unit *de-* of *detail* does not carry the reversative meaning 'do the opposite of' (MWO) and therefore cannot be considered a morph.

English negative morpheme realized by /ɪn/ in *inadequate* and /ɪm/ in *impossible*, the following regularity can be postulated:

- When followed by the **bilabial sounds** [p] and [b] (i.e. sounds that are articulated with both the lower and the upper lip), the negative morpheme is realized by the morph /ɪm/. E.g. *impossible* → /ɪmˈpɒsɪb(ə)l/, *imbalance* → /ɪmˈbæləns/.

- When followed by the sounds [m], [l], and [r], it is realized by /ɪ/. E.g. *immoral* → /ɪˈmɒrəl/, *illegal* → /ɪˈliːgəl/, *irretrievable* → /ɪrɪˈtriːvəb(ə)l/.

- In all other cases, it is realized by /ɪn/. E.g. *inadequate* → /ɪnˈædɪkwət/, *incompetent* → /ɪnˈkɒmpɪtənt/, *intolerable* → /ɪnˈtɒlərəb(ə)l/, etc.

To summarize: the morphs /ɪm/, /ɪ/, and /ɪn/ express the same negative meaning 'not' and never occur in the same environment: there can be no */ɪnˈpɒsɪb(ə)l/, */ɪmˈædɪkwət/, */ɪnˈmɒrəl/, etc. Accordingly, they are allomorphs of the same morpheme which occur in complementary distribution.

Finally, let us also consider the distribution of synonymous morphs which are capable of occurring in the same environment. Compare, for example, the negative morphs /ʌn/ of *untrue* and /ɪn/ of *inadequate*. From a **diachronic point of view** (i.e. if we consider the history of the English language), these morphs could perhaps be regarded as realizations of two different morphemes: /ʌn/ is a native Germanic morph which has existed since the **Old English period** (until ~1100), whereas /ɪn/ came into English only in the 14th century with loans from French such as *incombustible, incomprehensible, ineffectual*, etc. (For details, see Marchand 1969: 168-170 or the corresponding entries in the OED). By contrast, from a **synchronic point of view** (i.e. if we consider Present-day English only), the morphs /ʌn/ and /ɪn/ can be regarded as allomorphs of the same negative morpheme. This conclusion is supported by the fact that both /ʌn/ and /ɪn/ express the same negative meaning – *untrue* means 'not true' and *inadequate* means 'not adequate' – and occur in the same environment (i.e. before adjectives like *true* and *adequate*). It is of course true that in neither *untrue* nor *inadequate* is there a free variation of the morphs /ʌn/ and /ɪn/. For neither **intrue* nor **unadequate* exist in Present-day English.[6] However, we do find such adjective pairs as *uncommunicative* and *incommunicative* both meaning 'not communicative', *undistinguishable* and *indistinguishable* both

---

[6] The reason for this is that the syntactics of the morph *in-* allows it to combine only with words of Latin or Romanic origin, whereas the syntactics of *un-* allows it to be used only with native or completely naturalized words (OED). E.g. *true* is a word of Germanic origin and therefore combines with *un-*, whereas *adequate* is a word of Latin origin and therefore takes *in-*. The morphs *un-* of *untrue* and *in-* of *inadequate* can thus be regarded as allomorphs of the same morpheme which occur in complementary distribution.

meaning 'not distinguishable', *unmovable* and *immovable* both meaning 'not movable', etc. Hence, at least as far as these cases are concerned, we can speak of the free variation of the allomorphs /ʌn/ and /ɪn/.

As Marchand (1969: 170) points out, the morph *in-* is stylistically different from the morph *un-* in that it "forms learned, chiefly scientific words", whereas *un-* is a stylistically neutral regular negative morph. The difference between negative adjectives beginning with *un-* and *in-* is thus that *in*-adjectives sound more scientific and learned than their counterparts beginning with *un-*. Accordingly, the free variation of the morphs /ʌn/ and /ɪn/ in pairs like *uncommunicative* and *incommunicative* is an instance of stylistically relevant free variation.

Another interesting case involving synonymous morphs is represented by the distribution of *-ed* of *talked, walked, worked*, etc. and *ex-* of *ex-ambassador, ex-husband, ex-president*, etc. (Nida 1948: 425-426). If we again disregard the diachronic history of these signifiers, namely, that *-ed* is a native Germanic morph, whereas *ex-* has Latin origin (OED), we can arrive at the conclusion that *-ed* and *ex-* are realizations of the same past time morpheme. Indeed, a sentence like (45) indicates that the event in question took place before the moment of utterance (i.e. in the past).

(45)   He work<u>ed</u> with President Bush (COCA)

Similarly, *ex-* of *ex-president* indicates a past state, namely, that the person referred to as *ex-president* was president before the moment of utterance.

Now, let us also consider the distribution of the putative allomorphs *-ed* and *ex-*. The former occurs only after verbs (*talked, walked, worked*), but never before nouns (**ed-ambassador*, **ed-husband*, **ed-president*). Conversely, the form *ex-* occurs only before nouns (*ex-ambassador, ex-husband, ex-president*), but never after verbs (**talkex*, **walkex*, **workex*). Consequently, *-ed* and *ex-* are allomorphs of the same morpheme that occur in complementary distribution.

However, this conclusion seems to be counter-intuitive. Among other things, this impression emerges from the fact that this putative instance of complementary distribution is not **phonologically-conditioned**. That is, for example, in the case of *impossible*, there is a phonological explanation accounting for the fact that the negative morpheme is not realized by the morph /ɪn/. From the articulatory point of view, it is easier to say /ɪm'pɒsɪb(ə)l/ than */ɪn'pɒsɪb(ə)l/ because both [m] and [p] are bilabial sounds, whereas [n] is characterized by an alveolar articulation, i.e. its active articulator is the tip of the tongue, which is raised towards the alveolar ridge. The morph /ɪm/ is thus a product of the **place-of-articulation assimilation** of the alveolar [n] to the bilabial [p]. In contrast, no similar explanation can account for the impossibility of forms like **ed-ambassador* and **talkex*.

Nevertheless, the mere absence of phonological conditioning does not justify the treatment of -*ed* and *ex*- as realizations of two different morphemes. Consider, for instance, the plural form of the noun *ox*. In contrast to e.g. *box*, which takes one of the regular (phonologically conditioned) plural morphs /ɪz/ (*box* → *boxes*), *ox* combines with the morph /ən/: *ox* → *oxen*. It is obvious that this fact cannot be regarded as an instance of phonological conditioning: if speakers of English can pronounce /bɒksɪz/, they must also be able to pronounce /ɒksɪz/. (This is an instance of **lexical conditioning**, i.e. it is a peculiarity of the noun *ox* that it forms the plural form with the morph /ən/ rather than with /ɪz/. Similarly, it is a peculiarity of the noun *mouse* that its plural form is *mice*, not *mouses.) However, in spite of the non-phonological conditioning of *oxen*, the irregular morph /ən/ qualifies as an allomorph of the regular /ɪz/. Both express the same plural meaning and occur in complementary distribution: /ən/ is used with *ox*, whereas /ɪz/ attaches to the majority of other nouns ending in /s/. Accordingly, returning to the forms -*ed* and *ex*-, we must either grant them the status of allomorphs of the same past time morpheme which occur in non-phonologically conditioned complementary distribution or think of another explanation for why this cannot be the case.

With regard to the latter, it appears that our reluctance to regard -*ed* and *ex*- as allomorphs of the same morpheme also stems from the fact that these forms occur in **tactically different environments** (Nida 1948: 421), i.e. they combine with words belonging to different parts of speech. As stated above, -*ed* follows verbs, whereas *ex*- precedes nouns. (In contrast, all other allomorphs that have been dealt with so far occur in **tactically identical environments**. E.g. /ʌn/ and /ɪn/ precede adjectives; /ɪz/ and /ən/ follow nouns; etc.) However, the occurrence in tactically different environments does not suffice to deny -*ed* and *ex*- the status of allomorphs: the fact that they combine with words belonging to different parts of speech can be seen as an instance of complementary distribution. Being aware of this fact, Nida (1948: 425) proposes the following solution: -*ed* and *ex*- would qualify as allomorphic realizations of the same past time morpheme if English had another past time signifier possessing the distributional properties of both -*ed* and *ex*-, i.e. a form being able to occur both after verbs (e.g. *workeg meaning 'worked in the past') and before nouns (e.g. *eg-president meaning 'former president'). Since English lacks such a form, -*ed* and *ex*- are not allomorphs of the same morpheme but morphs realizing two different morphemes (which have identical signifieds).

Unfortunately, Nida does not really explain why the existence of such a form would suffice to regard -*ed* and *ex*- as allomorphs of the same morpheme. That is why it is not clear why the absence of such a form must suffice to deny -*ed* and *ex*- the status of allomorphs. Apart from this, the conclusion drawn by Nida is clearly at odds with what we said about the distribution of signifiers having the same signified: as we concluded in 2.2.5, two signifiers associated with the same signified are instances of the same sign (in our case, allomorphs of the same morpheme), regardless of their occurrence in tactically different environments.

Taking this into consideration, this book proposes a different solution. When analyzing the distribution of synonymous forms, which seem to express the same meaning, one can always try to find a slight semantic difference between them and thus be justified in regarding the forms in question as morphs realizing two different morphemes. This is a promising approach because, as has been pointed out by many authors, **absolute synonymy** (i.e. the existence of two forms which express exactly the same meaning) does not occur very often (see e.g. Cruse 2004: 154-155). As an illustration, let us, first of all, consider the distribution of the synonymous forms *ex-* and *former*: both carry the past time meaning and occur before nouns like *ambassador, husband,* and *president*. Are *ex-* and *former* of pairs like *ex-president–former president* allomorphs of the same morpheme which occur in free variation or morphs that have slightly different meanings and thus realize two different morphemes?

Since *ex-* and *former* are forms of different origin – *ex-* is Latin, *former* is Germanic (OED) – we can perhaps analogize the distribution of *ex-* and *former* to that of the negative morphs *un-* and *in-* and thus conclude that they are allomorphs of the same morpheme which occur in stylistically relevant free distribution. However, in contrast to *un-* and *in-*, *ex-* and *former* do seem to express slightly different meanings. As argued by the blogger David Goddard (http://tinyurl.com/5uwxlxc), "for a lot of people, [*ex*] gives off a somewhat negative feel, whereas [*former*] seems somehow more dignified, more amicable". In this connection, it is interesting to note that Google yields many more search results for *ex-President Bush* than it does for *former President Bush*.[7] In May 2011 there were approximately 273000 hits for *former President Bush* and approximately 3370000 hits for *ex-President Bush*. One explanation for this could be the just named semantic difference between *ex-* and *former*. If *ex-* does indeed "give off a somewhat negative feel", as suggested by David Goddard, then Internet users who use *ex-President Bush* instead of *former President Bush* do not only refer to the fact that George Bush was a U.S. president before the moment of utterance (i.e. which *ex-* literally means) but also (perhaps subconsciously) express their rather negative attitude towards him. In other words, *ex-president* does not only mean 'a person who was president before the moment of utterance' but 'a person who was president before the moment of utterance and whom the utterer of *ex-president* does not particularly like'. (We can say that the morph *ex-* has undergone what some linguists call the **subjectivisation of meaning** (see e.g. Stein and Wright 1995), i.e. the meaning of *ex-* has become 'subjective' in that it now carries a rather subjective speaker's

---

[7] I am very grateful to Bridgit Nelezen for drawing my attention to this fact.

attitude to the referent of the noun preceded by *ex-*.) Since George Bush, Jr. was a rather unpopular U.S. president, it is not surprising that *ex-President Bush* occurs more often than *former President Bush*.

If we are right with this hypothesis, we can conclude that *ex-* and *former* are morphs that express (slightly) different meanings and hence realize two different morphemes. Similarly, given that *-ed* of *worked* does not carry that 'somewhat negative feel' inherent in *ex-* of *ex-president*, we can conclude that *-ed* of *worked* and *ex-* of *ex-president* express (slightly) different meanings and, accordingly, are not allomorphs of the same morpheme but morphs realizing two different morphemes.

In summary: we have become acquainted with six main types of the distribution of morphs in English:

1. **Contrastive distribution of two non-identical morphs expressing non-identical meanings**. The morphs in question are realizations of two different morphemes. E.g. the morphs /truː/ and /ˈeɪ.bl/ realizing the morphemes {true} and {able}.

2. **Contrastive distribution of two identical morphs expressing non-identical meanings**. The morphs in question are realizations of two different morphemes. E.g. the morphs /truː/ and /truː/ realizing the morphemes {true$_1$} 'faithful' and {true$_2$} 'not false'.

3. **Stylistically irrelevant free variation of two non-identical morphs expressing the same meaning**. The morphs in question are allomorphs of the same morpheme. E.g. /ˈdiːteɪl/ and /dɪˈteɪl/ realizing the morpheme {detail}.

4. **Stylistically relevant free variation of two non-identical morphs expressing the same meaning**. The morphs in question are allomorphs of the same morpheme. E.g. the morphs /ʌn/ and /ɪn/ realizing the same negative morpheme in *uncommunicative* and *incommunicative*.

5. **Phonologically conditioned complementary distribution of two non-identical morphs expressing the same meaning**. The morphs in question are allomorphs of the same morpheme. E.g. the morphs /ɪn/, /ɪm/, and /ɪ/ realizing the same negative morpheme in *inadequate*, *impossible*, and *illiterate*.

6. **Non-phonologically conditioned complementary distribution of two non-identical morphs expressing the same meaning**. The morphs in question

are allomorphs of the same morpheme. E.g. the morphs /ən/ and /ɪz/ realizing the same plural morpheme in *oxen* and *boxes*.

2.4 The segmentation of words into morphemes

As we established in Section 2.1, the word *untrue* can be segmented into the morphemes {un} and {true} because the meaning 'untrue' can be segmented into the meanings 'not' and 'faithful', which the components *un-* and *true* express in other environments: *un-* carries the meaning 'not' in e.g. *unable* and *true* carries the meaning 'faithful' when used without *un-*, i.e. in sentences like *Her lover had been true*. The word *untrue* is thus an instance of **isomorphism of formal and semantic segmentation** (Plungian 2000: 39). That is, the formal segmentation of the signifier *untrue* into the morphemes {un} and {true} is paralleled by the semantic segmentation of the signified 'not faithful' into the meanings 'not' and 'faithful', inherent in the morphemes {un} and {true}. The same applies to *unable*. The formal segmentation of the signifier *unable* into the morphemes {un} and {able} is paralleled by the semantic segmentation of the signified 'not able' into the meanings 'not' and 'able', inherent in the morphemes {un} and {able}. Isomorphic forms like *untrue* and *unable* present no theoretical difficulties and will therefore be largely left out of consideration in the remainder of this section.

2.4.1 Anisomorphism. Full-idiomaticity

One manifestation of **anisomorphism of formal and semantic segmentation** (which is the reverse of isomorphism) is full-idiomaticity. As was pointed out in 1.2.2, fully-idiomatic words include semantically modified words like *boyfriend*, whose idiomatic meanings are still explainable in terms of their components' literal meanings (recall that boyfriends are typically friends and former young male children), and opaque words like *understand*, whose idiomatic meanings are no longer transparent.

Unlike the isomorphic words *untrue* and *unable*, the idiomatic words *boyfriend* and *understand* raise the following questions. Can *boyfriend* be segmented into the morphemes {boy} and {friend}, even if *boyfriend* does not mean 'a friend who is a boy'? Similarly: can *understand* be segmented into the morphemes {under} and {stand}, even if *understand* does not mean 'to stand under somebody or something' but 'to grasp the meaning or the reasonableness of something'? In the following we will become acquainted with three possible theoretical solutions.

2.4.2 A purely semantic approach

If we want to adhere to the definition of a morpheme as a meaning-carrying unit, the answer to the questions raised above can only be 'no'. *Boyfriend* cannot be segmented into the morphemes {boy} and {friend} because the putative components *boy* and *friend* do not carry the meanings 'boy' and 'friend', which they carry in other environments. Likewise, *understand* cannot be segmented into the morphemes {under} and {stand} because the putative components *under* and *stand* do not carry the meanings 'under' and 'to stand', which they carry in other environments. *Boyfriend* and *understand* are monomorphemic words consisting of the morphemes {boyfriend} and {understand}, which carry the idiomatic meanings 'a male sexual partner of almost any age over puberty' and 'to grasp the meaning or the reasonableness of something'.

This solution, however, does not seem to be very convincing. Especially in the case of *understand*, the purely semantic approach is at odds with the fact that the verb under analysis is **headed** by the component *stand*. In morphology, the term 'head' refers to the most important structural component of a complex word, which determines its word-class as well as the **inflectional marking** (i.e. bound morphs expressing obligatory grammatical meanings such as e.g. 'the past tense'). Thus *understand* is a verb because it is headed by the verb *stand*, not by the preposition *under*. Also, the past tense of *understand* is not *understanded but *understood*. This is because the past tense form of the head element *stand* is also *stood*, not *standed. Both these facts do not support the analysis of *understand* as a monomorphemic word. (For similar observations, see Aronoff 1976: 14-15.)

2.4.3 Nida's purely formal approach

An example of a purely formal approach is Nida (1974: 58-59), who argues that a morpheme can be isolated if it satisfies one of the following conditions:

1. It occurs "in isolation".

2. It occurs "in multiple combinations in at least one of which the unit with which it is combined occurs in isolation or in other combinations".

3. It occurs "in a single combination provided that the element with which it is combined occurs in isolation or in other combinations with other non-unique constituents".

'In isolation' means that the signifier in question can form a one-word elliptical sentence and thus fulfills the isolatability criterion which we discussed in the previous chapter. Consider, for example, the underlined forms in (46), (47), and (48).

(46) Here, in a piece called 'The Problem With Boys,' Tom Chiarella broods, 'There is something odd and forbidden about the word boy. Typing it feels a little creepy, almost pornographic. <u>Boy</u>. A little word, naked and weak, an iconic expression of smallness, of vulnerability.... (COCA)

(47) Maleus found his cap on the chair arm and traced a finger around its rim. <u>Friend</u>? 'Oh, there was...' He shook his head (COCA)

(48) Some of our friends trotted up to see my beautiful babe, stuck their heads through the curtains. They tossed their heads, chortled and nibbled the back of her neck. 'Come on, little one. <u>Stand</u>! <u>Stand</u>!' (COCA)

As we can see, the signifiers *boy*, *friend*, and *stand* can form elliptical sentences and, accordingly, are isolatable as morphemes (or to be more precise, as morphs realizing morphemes) in accordance with Nida's Condition 1: it does not matter that the meaning 'boyfriend' does not contain the meanings 'boy' and 'friend' and the meaning 'understand' does not contain the meaning 'to stand'. As was already mentioned in 2.1, morphemes like {boy} and {friend} of *boyfriend* and {stand} of *understand* whose morphs are capable of occurring in isolation are traditionally called **free morphemes**.

Condition 2 is concerned with **bound morphemes**, i.e. morphemes whose morphs, in contrast to those of free morphemes, never occur in isolation. Consider, for example, the component *under* of *understand*. Despite the fact that the preposition *under* is usually separated from other signifiers by means of a blank space (e.g. *under the table*, not **underthetable*), *under* does not seem to be capable of occurring in isolation. Perhaps one could imagine a hesitant speaker who says *Under.* as a response to a question like *Where is my bag?* uttered by another speaker. However, neither COCA nor BYU-BNC contain elliptical sentences made up of *under* only. The same applies to other prepositions. For instance, as Haspelmath (2011: 40) points out, in the PP *to Lagos*, the preposition *to* is a bound form: it cannot occur on its own without something following it. Indeed, as in the case of *under*, neither COCA nor BYU-BNC contain elliptical sentences consisting only of the signifier *to*.

What follows from this is that the component *under-* of *understand* does not fulfill Bloomfield's isolatability criterion and therefore does not qualify as a word. However, it does fulfill Nida's Condition 2: *under-* of *understand* occurs in multiple combinations – e.g. *understand*, *under the table*, *under that*, etc. – in at least one of which (e.g. *understand*) the unit with which it is combined (i.e. *stand*) occurs in isolation. In other words, *under-* of *understand* is isolatable as a morpheme because in the word under analysis it combines with the unit *stand*, which occurs in isolation. Again, it does not matter that the literal meaning 'under' is not part of the meaning 'to understand'.

A similar example of a bound morph is the negative morph *un-* of *untrue*. Like *under-* of *understand*, *un-* of *untrue* is also not capable of forming elliptical sentences consisting of *un-* only. (Its bound character is more obvious than that of *under-* because in contrast to the latter, the former is usually not separated from other morphs by means of a blank space.) As in the case of *under-* of *understand*, *un-* of *untrue* is also isolatable as a morph realizing a morpheme in accordance with Nida's Condition 2: it occurs in multiple combinations (e.g. *untrue, unable*) in at least one of which (e.g. *untrue*) the unit with which it is combined (i.e. *true*) occurs in isolation: both COCA and BYU-BNC contain elliptical sentences made up of *true* only.

Besides bound morphs like *under-* of *understand* and *un-* of *untrue* which occur in combination with free morphs, there are bound morphs which occur in other combinations. Consider, for instance, the verb *receive*. Applying the latter part of Condition 2, we can segment *receive* into the morphemes {re} and {ceive}, even though the meaning 'receive' cannot be segmented into two independent meanings attributable to the putative morphs *re-* and *-ceive*. The unit *re-* qualifies as a morph of a morpheme because in addition to occurring in *receive*, it also occurs in *reduce, refer,* and *retain*. Likewise, the unit *-ceive* qualifies as a morph of a morpheme because in addition to occurring in *receive*, it also occurs in the verbs *conceive, deceive,* and *perceive*.

Finally, Nida's Condition 3 helps us deal with **unique morphemes**, i.e. morphemes whose morphs occur only in one particular environment. An often cited example is *cran-* of *cranberry*. As argued by many authors (e.g. Aronoff 1976: 10; Taylor 2002: 273), *cran-* is a unique morph because it occurs only in combination with the free morph *berry* in the word *cranberry*. Strictly speaking, this is no longer true of the English language: the unit *cran* does occur as an elliptical version of *cranberry* in *cranapple juice, cranapple crunch, cranapple pie,* and the like. E.g. (49).

(49) <u>Cran</u>apple Juice – A drink produced by squeezing or crushing <u>cran</u>apples (http://www.drinkswap.com/cranapple-juice.htm)

Anyway, the point of Nida's Condition 3 is that bound unique units can be isolated as morphemes when they occur in combination with free morphs. Thus if *cran* of *cranberry* still remained a unique bound unit, it would nevertheless qualify as a morph realizing a morpheme because in the word *cranberry* it combines with the unit *berry*, which, like other nouns, is capable of occurring in isolation.

2.4.4 Nida's approach and the conception of differential meaning

The major problem with Nida's formal approach seems to be the fact that it grants the morphemic status not only to opaque units like *under-* and *stand* of *understand* but also to entirely meaningless *re-* and *-ceive* of *receive*. Indeed, if *re-* and *-ceive* are isolatable as morphemes even though they do not mean anything (but only occur in other combinations), a question arises as to whether we can preserve the definition of a morpheme as a meaning-carrying unit. That is, if units that do not carry any meanings can nevertheless be regarded as (morphs realizing) morphemes, can a morpheme be still regarded and defined as a meaning-carrying unit?

According to Ginzburg et al. (1979: 24), Nida's formal approach can be reconciled with the classic conception of a morpheme as a meaning-carrying unit if we assume that opaque components which can be isolated as morphs from a formal point of view only possess the **differential meaning**. As an illustration, let us again consider *understand*. As was stated in 1.2.2, its putative components *under-* and *stand* are opaque because the idiomatic meaning 'to understand' can hardly be accounted for in terms of the meanings 'under' and 'to stand', which *under-* and *stand* express in other environments. However, regardless of this fact, both these units can be said to carry differential meanings. This means that the unit *under-* contributes to the overall meaning of *understand* by making *understand* distinguishable from e.g. *withstand*. And the unit *stand* contributes to the overall meaning of *understand* by making *understand* distinguishable from e.g. *undergo*. In a similar way, the meaningless units *re-* and *-ceive* of *receive* can be said to possess differential meanings because they make *receive* distinguishable from words like *conceive*, *deceive*, *perceive*, etc. and *reduce*, *refer*, *retain*, etc. which likewise contain either the empty unit *re-* or the empty unit *-ceive*.

The conception of differential meaning is, however, also far from being unproblematic. If we accept the view that 'differential meaning' is a sufficient morpheme-hood condition (i.e. the minimum requirement that the form in question must fulfill in order to qualify as a morpheme), then almost any combination of sounds or even individual sounds become isolatable as morphemes. For example, we can argue that the word *car* consists of the morphemes {c} and {ar}: the unit *c-* qualifies as a morpheme because it makes *car* distinguishable from *bar*. And the unit *-ar* is a morpheme because it makes *car* distinguishable from *cable*. But in this case, we would no longer be able to distinguish between morphemes (meaning-carrying units) and phonemes (meaning-distinguishing units), since any sound capable of distinguishing meaning (e.g. /k/ of *car*), which has been traditionally regarded as (a phone realizing) a phoneme, automatically becomes (a morph realizing) a morpheme.

2.4.5 Mel'čuk's theory of quasi-linguistic units

Another attempt to reconcile Nida's formal approach with the traditional conception of a morpheme as a meaning-carrying unit is Mel'čuk's theory of **quasi-linguistic units** (2001: 278-285). According to Mel'čuk, quasi-linguistic units are forms that are isolatable as morphemes from a formal but not from a semantic point of view. For example, as we established in 2.4.3, the words *boyfriend* and *understand* can be segmented into the morphemes {boy} / {friend} and {under} / {stand}: the components *boy*, *friend*, and *stand* are free morphs, which occur in isolation, and the component *under-* is a bound morph which occurs in multiple combinations with free units like *stand*. However, *boyfriend* does not mean 'a friend who is a boy' and *understand* does not mean 'to stand under somebody or something'. Accordingly, the units *boy* / *friend* and *under-* / *stand* are not morphs realizing morphemes but quasi-linguistic units (or **quasi-signs**).

A very important aspect of Mel'čuk's approach is the classification of quasi-linguistic units into **morfoids** and **submorphs**. Morfoids are units that occur in semantically motivated idiomatic words like *boyfriend*, whereas opaque words like *understand* and words like *receive* which are segmentable into meaningless units occurring in other combinations are said to consist of submorphs.

As Mel'čuk acknowledges, sometimes it is rather difficult to decide in favor of either the morfoid or the submorph solution, since one and the same idiomatic form can be perceived as fully-motivated by one speaker and as opaque by a different speaker. However, as will be shown below, this fact must be seen not as a shortcoming but as one of the major advantages of this approach. Consider, for example, the fully-idiomatic adjective *bananas* 'crazy' (MWO), which often occurs in the fully-idiomatic VPs *go bananas* and *drive bananas*. Similar to our previous examples *boyfriend* and *understand*, *bananas* seems to be formally segmentable into the free morph *banana* and the bound morph *-s*. However, the meaning 'bananas' does not contain the meanings 'banana' and 'plurality', which the components *banana* and *-s* express in other environments. Accordingly, the units *banana* and *-s* are not morphs realizing morphemes but quasi-linguistic units.

As regards the question of whether the components *banana* and *-s* are morfoids or submorphs, it appears that the fully-idiomatic meaning 'crazy' has nothing in common with the meanings 'banana' and 'plurality'. Accordingly, one may be tempted to conclude that *banana* and *-s* are opaque submorphs rather than motivated morfoids. However, as argued by the author of the Web page http://www.englishdaily626.com/slang.php?054,

> when apes are given a bunch of bananas, they eat them with tremendous enthusiasm, as though they've lost their minds.

This is most likely a **popular** (or **folk**) **etymology** of *bananas*, i.e. an explanation invented by some particularly creative speaker of English who wanted to find at least some connection between the idiomatic meaning 'crazy' and the literal meaning 'more than one banana'. Nevertheless, this explanation can be used as a justification for the treatment of the components *banana* and *-s* as morfoids: for those speakers of English who do indeed attribute the idiomatic meaning 'crazy' of *bananas* to the craziness of apes caused by eating more than one banana, the word *bananas* does indeed consist of the two morfoids *banana* and *-s*, whose literal meanings 'banana' and 'the plural number' partially motivate the idiomatic meaning 'crazy'.

The major advantage of Mel'čukian approach is thus that it allows us to distinguish between different groups of speakers (of English or any other language) who may have different judgments as to whether the idiomatic meaning under analysis is opaque or is somehow motivated by its components' literal meanings. In the former case, the word under consideration becomes segmentable into submorphs; in the latter case into morfoids.

Note that the theory of quasi-linguistic units does not claim that any word can be segmented into at least two submorphs. Thus, given the segmentation of *receive* into the submorphs *re-* and *-ceive*, why can we not segment e.g. *car* into the submorphs *c-* and *-ar*? As stated in 2.4.4, both these units occur in other combinations – *c-* also occurs in *cable* and *-ar* also occurs in *bar* – and therefore seem to qualify as submorphs. However, the fact that *c-* occurs not only in *car* but also in *cable* and *-ar* occurs not only in *car* but also in *bar* seems to be a matter of coincidence. The members of the pairs *car* / *cable* and *car* / *bar* do not have much in common. That is why the segmentation of *car* into the units *c-* and *-ar* does not seem natural. In contrast, the segmentation of *receive* into the units *re-* and *-ceive* does not seem unnatural because *-ceive* makes *receive* distinguishable from a semantically related word *reception*. Compare also *deceive* and *deception*, *reduce* and *reduction*, and examples alike.

2.4.6 Anisomorphism. Partial idiomaticity

Having established how to deal with fully-idiomatic forms like *boyfriend* and *understand*, we can now proceed to the morphemic analysis of words which are partially idiomatic. We will begin with *blackboard*.

As we said in 1.2.2, the meaning 'blackboard' contains the meaning of the component *board* but not of the component *black*. A blackboard is not a black board but a board for drawing or writing upon with chalk. Accordingly, of its two components, which are isolatable as morphemes from a formal point of view, only *board* can be regarded as a morph realizing a morpheme: it can occur in isolation and its meaning is part of the meaning 'blackboard'. By contrast, *black* is a morfoid whose literal meaning only motivates the idiomatic meaning 'board for writing': blackboards have dark surfaces and the black color is the most prominent representative of dark colors.

Consider also the morphemic structure of *twilight*. From a formal point of view, *twilight* is segmentable into the components *light* and *twi-*: *light* is a free morph, which can occur in isolation, whereas *twi-* is a non-unique bound unit, which occurs in multiple combinations with forms occurring in isolation. For example, in addition to occurring in *twilight*, *twi-* also occurs in e.g. *twi-headed* (meaning 'having two heads'.) However, from a semantic point of view, the morphemic status can only be granted to the component *light*. Thus, according to MWO, *twilight* does not mean 'two lights' or 'double light' but

> the light from the sky between full night and sunrise or between sunset and full night produced by diffusion of sunlight through the atmosphere and its dust. (MWO)

It follows that the component *twi-* is not a morph but a quasi-linguistic unit.

With regard to the question of whether *twi-* of *twilight* is a morfoid or a submorph, it appears that this unit, in contrast to *black* of *blackboard*, is semantically opaque : it is not clear what the sense 'twilight' has in common with the meaning 'two, double', which *twi-* expresses in other environments. As just pointed out, *twilight* does not mean 'two lights' or 'double light' but 'the light from the sky between full night and sunrise'. Accordingly, the component *twi-* of *twilight* is not a morfoid but a submorph.

In summary, semi-idiomatic words can consist of:

1. **one normal morph and one motivated morfoid** (e.g. *blackboard*)
2. **one normal morph and one opaque submorph** (e.g. *twilight*)

As in the case of fully-idiomatic words, two different speakers may have different judgments regarding transparency / opacity of the idiomatic component of the semi-idiomatic word under analysis. In this case, one and the same semi-idiomatic word becomes segmentable into one normal morph and one motivated morfoid (this applies to the speaker for whom the idiomatic component is transparent) and one normal morph and one opaque submorph (this applies to the speaker for whom the idiomatic component is opaque).

2.4.7 Anisomorphism. Additional meanings

Finally, we will consider anisomorphic words whose meanings contain not only their components' literal meanings but also additional, unpredictable idiomatic meanings. Let us begin with the word *football*.

From a formal point of view, *football* is easily segmentable into the components *foot* and *ball*: both can occur in isolation and thus qualify as free morphs in accordance with Nida's Condition 1. However, *football* does not mean 'a ball for a foot' (which its components literally stand for), but 'a particular sport that involves a ball and the players' feet trying to kick it to the goal at the opposite side of the field' (OED). In other words, the meaning 'football' contains the meanings of the components *foot* and *ball* – these components provide a (somewhat incomplete) explanation for how football is played – but in addition to these meanings, it also contains the idiomatic meaning 'a particular sport different from basketball, handball, etc.', this meaning being inherent in neither *foot* nor *ball*. Obviously, words like *handball, basketball, volleyball*, etc. exhibit a similar semantic structure. Their meanings contain their components' literal meanings, which provide an incomplete explanation for how these games are supposed to be played – e.g. handball is the game that involves a ball and the players' hands throwing it to the goal at the opposite side of the field – but in addition to these meanings, they also contain the idiomatic meaning 'a particular sport different from other sports'.

Consider also the meaning of *stealer* in (50).

(50)   *But more than that, he feels guilty. Because now he is a thief, and not just any thief, but a <u>stealer</u> of dreams and wishes* (COCA)

As argued in some studies, the word *stealer* does not exist in English because the meaning 'a person who steals' is expressed by the signifier *thief*. However, as (50) demonstrates, this is not so. According to the OED, the noun *stealer* appeared in the English language in 1508. It originally meant 'one who steals; a thief' (and thus used to be a full-synonym of *thief*) but with the course of time has undergone semantic narrowing and, as a result, come to mean 'one who steals something specified': Example (50), in which *stealer* is contrasted with *thief*, is a good illustration of the fact that in Present-day English the meaning 'stealer' is indeed narrower than the meaning 'thief'.

A similar case is the noun *writer* of (51).

(51)   *A prolific <u>writer</u>, he has published three short story collections, 12 picture books (including the highly acclaimed Zoom trilogy) and three young adult novels* (COCA)

At first glance, it may seem that the meaning of the *writer* of (51) is made up of the meanings of its components: the free unit *write*, which is capable of occurring in isolation, and the non-unique bound unit *-er*, which occurs in multiple combinations with units occurring in isolation; e.g. *browser, preacher, stealer, teacher*. That is, *writer* means 'a person who writes, a performer of the action of writing'. However, not any person who writes qualifies as a writer. For instance, a person who is writing an e-mail message can hardly be referred to as a writer. (And a person who has written a lot of e-mail messages definitely cannot be referred to as a prolific writer.) A writer is a person who produces particular products of writing: story collections, picture books, adult novels, and the like. In other words, in addition to the meanings 'to write' and 'performer of some action', inherent in the components *write* and *-er*, the meaning 'writer' contains an additional idiomatic meaning 'particular products of writing', which cannot be attributed to either *write* or *-er*.

Recall also the formation *waithood*, which we discussed in Chapter 1. As we concluded in 1.2.2, the meaning 'waithood' does not only contain the meanings 'to wait' and 'stage', inherent in the components *wait* and *-hood*, but also the idiomatic meaning 'a particular waiting stage involving a colleague graduate who is waiting for a good job and / or financial security in his life and therefore postpones marrying'.

Given these semantic structures of the words *football, stealer, writer*, and *waithood*, we need to answer the question of where the idiomatic meanings 'a particular sport' of *football*, 'specified objects of stealing' of *stealer*, 'particular products of writing' of *writer*, and 'a particular waiting stage' of *waithood* actually come from. We do not need to be concerned with the question of whether these words can be segmented into the morphemes {foot} / {ball}, {steal} / {er}, {write} / {er}, and {wait} / {hood}. The answer to this question is obvious: if the meanings 'football', 'stealer', 'writer', and 'waithood' contain the meanings of these components, then all of them qualify as morphs realizing normal morphemes.

To account for the presence of the additional idiomatic meanings in *football, stealer, writer*, and *waithood*, we can resort to either the **mega-morph** or the **zero morph approach**. In the former case, we attribute the idiomatic meaning in question to one of the two overt components of the word under analysis. This component then becomes a **mega-morph** (Mel'čuk 2001: Ch. 7), that is, a morph which **cumulatively expresses** more than one signified. For instance, in the case of *stealer*, we can conjecture that the idiomatic meaning 'specified objects of stealing' is inherent in the morph *-er*, which cumulatively expresses this meaning together with the literal meaning 'performer of some action'. We are justified in arriving at this conclusion because in English there is also the semantically and formally related noun *writer*, whose signified does not only contain the meanings 'to write' and 'performer of some action' (i.e. the literal meanings of the overt components *write* and *-er*) but also the idiomatic meaning 'particular products of writing'. Similarly, it appears that in *football* the idiomatic meaning 'a particular sport' is inherent in the morph *ball* rather than in *foot*. This is because in addition to *football*, there are the words *handball, basketball, volleyball*, etc., whose signifieds exhibit the same idiomatic pattern as *football*.

An alternative to the mega-morph approach is the zero morph approach, i.e. the attribution of an idiomatic meaning to a **zero morpheme**, i.e. a morpheme that does not have an overt morph. For instance, instead of attributing the idiomatic meaning 'specified objects of stealing' to the morph *-er*, we can segment the word *stealer* into:

- the overt morph *steal* carrying the meaning 'to steal'
- the overt morph *-er* carrying the meaning 'performer of some action'
- a covert zero morph {ø} carrying the idiomatic meaning 'specified objects of stealing'.

Similarly:

- *writer* = {write} + {er} + {ø}, in which {ø} is a covert zero morph carrying the idiomatic meaning 'particular products of writing'.

- *football* = {foot} + {ball} + {ø}, in which {ø} is a covert zero morph carrying the idiomatic meaning 'a particular sport'.

- *waithood* = {wait} + {hood} + {ø}, in which {ø} is a covert zero morph carrying the idiomatic meaning 'a particular waiting stage'.

This textbook rejects both these approaches. The reason for this is that both the mega-morph approach and the zero-morph approach are based on what Plungian (2000: 39) calls the **additive principle of morphology**, i.e. a model describing an ideal signifier–signified relation. In accordance with this model, if Signified A can be segmented into independent meanings X and Y, then A's signifier must likewise be segmentable into signifiers {x} and {y} carrying meanings X and Y.

The additive model works perfectly with isomorphic words like *untrue*. That is, we have the impression that the signified 'untrue' is a semantically complex signified segmentable into two independent meanings: 'not' and 'faithful'. Given this and given that *un-* and *true* express the meanings 'not' and 'faithful' in other environments, we can easily segment *untrue* into the morphs *un-* and *true*. By contrast, anisomorphic words like *stealer* are perceived as deviations from the additive model. Thus, while the signified 'stealer' is segmentable into three independent meanings 'to steal', 'performer of some action', and 'specified objects of stealing', the signifier *stealer* can be segmented only into two morphs:

steal and *-er*. Similarly, while the signified 'waithood' is segmentable into three independent meanings 'to wait', 'stage', and 'a particular waiting stage', the signifier *waithood* is segmentable only into two morphs: *wait* and *-hood*. To account for these deviations, we then need to attribute the idiomatic meanings 'specified objects of stealing' and 'a particular waiting stage' to either one of their overt components, which then becomes a mega-morph, or to a covert zero morph.

But is this really necessary? Do anisomorphic words like *stealer* and *waithood* represent deviations from the 'normal' signifier–signified relation exemplified by fully-isomorphic words like *untrue*? The answer to this question is of course 'no'. As has been repeatedly pointed out by many authors (especially those who study idiomatic meanings), fully-isomorphic complex words like *untrue* whose signifieds are representable only in terms of their components' signifieds are considerably outnumbered by anisomorphic words like *stealer* and *waithood* whose meanings contain not only their components' literal meanings but also additional, unpredictable idiomatic meanings. In other words, the default case is represented by an anisomorphic complex word like *stealer* rather than by an isomorphic complex word like *untrue*. (This will become especially obvious in Chapters 4 and 5.)

Taking this into account, we can conclude this section with the claim that the anisomorphic words *football*, *stealer*, *writer*, *waithood*, etc., whose signifieds contain additional idiomatic meanings, must be segmented only into two normal morphemes, i.e. the morphemes {foot} and {ball} of *football*, {steal} and {er} of *stealer*, {write} and {er} of *writer*, and {wait} and {hood} of *waithood*. Given the default character of the signifier–signified relation exemplified by these words, there is no need to account for additional idiomatic meanings by means of postulating mega-morphs or zero morphs.

2.5 The hierarchy of morphs and units alike

The last section of this chapter deals with the hierarchical differences between different morphs (as well as morfoids and submorphs) into which complex forms can be segmented.

As the starting point, let us consider the morphs *un-* and *true* of *untrue*. It appears that one of these morphs – the morph *true* – is both formally and semantically more independent than the morph *un-*. From a formal perspective, *true* is a free morph, which can occur in isolation, whereas *un-* is a bound morph which occurs in combination with free morphs like *true* and *able*. Similarly, from a semantic perspective, the meaning 'true', inherent in the morph *true*, seems to be a more independent meaning than the negative meaning inherent in

The internal structure of English words

the morph *un-*: while the former can be expressed by *true* alone, the latter requires a combination of *un-* with free morphs like *true* and *able*.

Traditionally, this contrast between morphs like *un-* and *true* of *untrue* has been captured with the help of the **affix–root** dichotomy. That is, the morph *true* is the root of the word *untrue*, whereas the morph *un-* has the status of an affix.

2.5.1 Affixes versus roots

The most important differences between roots and affixes are as follows.

- Affixes are bound units; roots can be both free and bound.

- Affixes are shorter than roots.

- Roots are obligatory elements; affixes are optional elements.

Now, let us briefly discuss each of these differences. The defining characteristic of an affix is that it is a bound unit, which occurs only in combination with other units. In contrast, a root can be represented by both a free and a bound element. Prototypical roots are free units like *true* of *untrue*, *foot* and *ball* of *football*, etc., but bound roots can be found as well.

Consider, for instance, the noun *history*. From a formal point of view, *history* can be segmented into the components *histor-* and *-y*. Both of them occur in other combinations: the unit *histor-* also occurs in *historical* and the unit *-y* also occurs in *theory*. (The word *theory* is thus likewise segmentable into the components *theor-* and *-y*.) With regard to the semantic perspective, it is clear that the meaning 'history' cannot be segmented into two independent meanings attributable to the putative morphs *histor-* and *-y*. Like *re-* and *-ceive* of *receive*, *histor-* and *-y* of *history* do not have discernible meanings of their own and thus qualify as quasi-linguistic units of the submorphic type.

With regard to the hierarchy of the submorphs *histor-* and *-y*, it is intuitively clear that the unit *histor-* is more like a root, whereas *-y* is more like an affix. This impression arises, however, not because of the free–bound distinction between them – both *histor-* and *-y* are bound submorphs, which never occur in isolation – but because the submorph *histor-* is longer than the submorph *-y*: whereas the latter is made up of only one sound /ɪ/, the former consists of six sounds: /ˈhɪstər/.

The second often named difference between roots and affixes is thus that the latter are typically shorter than the former. In English there are affixes that consist of only one sound. For example, there are affixes like /ɪ/ of *cloudy*, *rainy*, etc. (which is semantically unrelated to the /ɪ/ of *history* and *theory*) or /æ/ of *asymmetric* /æsɪˈmɛtrɪk/, *asymbolic* /æsɪmˈbɒlɪk/, etc. In contrast, the shortest English root consists of two sounds: /aɪ/ of *eye* and *I*. Hence, concluding our analysis of *history*, we can argue that this word consists of the bound root *histor-* and the affix *-y* because the latter is much shorter than the former.[8]

Third, while roots are obligatory elements, affixes are optional. That is, a word must contain at least one root, but it may contain no affixes. For example, the monomorphemic words *table, boy, friend, dog, cat*, etc. contain no affixes. Accordingly, there can be no words consisting of affixes only. That is why words like *history* cannot be segmented into two bound affixes: *histor-* and *-y*. Given the general obligatoriness of roots, one of these units must be necessarily regarded as a root.

Finally, it is sometimes argued that roots are different from affixes with regard to their signifieds: roots usually express more concrete signifieds than affixes. As an illustration of this, let us consider the word *morphology*. From a formal point of view, *morphology* can be segmented into three units: *morph, -o-*, and *-logy*. The unit *morph* is a free unit, which is capable of occurring in isolation, and thus can be regarded as a root of *morphology*. The unit *-o-* is a non-unique bound unit, which occurs in combination with other free units. E.g. *phonology*. Since the unit *-o-* consists of only one sound, it can be regarded as an affix. Finally, the unit *-logy* is a non-unique bound unit which occurs in words like *morphology, phonology, lexicology*, etc., where it carries the meaning 'names of sciences, departments of study' (OED): whereas morphology is the branch of linguistics that is concerned with morphs, phonology can be defined as the study of phones; lexicology is the study of the lexis; and so on. (The word *morphology* can thus be segmented into the morphs *morph* and *-logy* and the meaningless submorph *-o-*, which serves as an interlinking element between them.) The meaning 'study', expressed by the morph *-logy*, seems to be a more independent and less abstract meaning than e.g. the meaning 'negation', inherent in the morph *un-* of *untrue*. Whereas the former can be easily paraphrased as the acquisition of some knowledge, the latter represents one of the most difficult and hardly definable concepts. Given this fact, some authors conclude that despite its bound status, the morph *-logy* is a root rather than an affix.

This textbook rejects the categorical analysis of *-logy* of *morphology* as a root, especially if this analysis takes into account only the 'not so abstract' nature of the signified inherent in *-logy*. It is of course true that in contrast to the formal categories which we discussed in the previous chapter (i.e. the word, the phrase, the clause, and the sentence), the morpheme is both a formal and a semantic category. However, the distinction between the root and the affix seems to be a formal rather than a semantic distinction: as stated above, the defining characteristic of an affix is that it is a bound morph, which cannot occur in

---

[8] In a similar way, we can argue that *receive* consists of the bound root *-ceive* and the affix *re-* because the latter is shorter than the former.

The internal structure of English words 53

isolation. Accordingly, roots must be distinguishable from affixes only on the basis of formal, not semantic criteria. With regard to the former, there is no special reason to regard *-logy* as a root. First of all, the morph *-logy* meaning 'names of sciences, departments of study' never occurs in isolation. Second, in English there are affixes which, like *-logy* /ləd͡ʒɪ/, consist of five sounds: for example, the adjective-building affix *-aceous* /ˈeɪʃəs/ of words like *curvaceous*, *rudaceous*, etc. Finally, the analysis of *-logy* as an affix does not violate the obligatoriness requirement: the word *morphology* has the root *morph*, so that both the units *-o-* and *-logy* can be regarded as affixes.

2.5.2 Combining form as a distinct morpheme type?

Morphs like *-logy* of *morphology* that have the formal properties of affixes and the semantic properties of roots are often called **combining forms**. According to the OED, in Present-day English there are 2179 combining forms. These include, for example:

- *-babble* of nouns like *edubabble, neurobabble, designer-babble*, etc. which denote 'various types of confusing or pretentious jargon, esp. that characteristic of a specified field or group' (OED).

- *-core* of nouns like *emo-core, noisecore, sadcore*, etc. which designate '(usually more extreme or intense) subgenres of popular music, esp. punk, grunge, techno, and heavy metal' (OED).

- *e-* of nouns like *e-commerce, e-journal, e-signature*, etc. which denote 'involvement in electronic media and telecommunications (esp. the use of electronic data transfer over the Internet, etc.), usually to distinguish objects or actions from their non-electronic counterparts' (OED).

- *-erati* of nouns like *belligerati, chatterati, geekerati*, etc. which designate 'elite or prominent groups of people who are associated with what is specified by the stem word' (OED).

- *-tainment* of nouns like *docutainment, edutainment, irritainment*, etc. which denote 'genres of broadcasting, journalism, etc., in which entertainment is combined with aspects of the genre, etc., indicated by the first element' (OED).

As in the case of *-logy* of *morphology*, the answer to the question of whether a particular combining form can be regarded as a root or an affix must depend only on the fulfillment of the formal criteria which were introduced in 2.5.1. That is, for example, despite the fact that *e-* of *e-commerce, e-journal, e-signature*, etc. is associated with a 'not so abstract' meaning 'electronic', the combining form under analysis must be regarded as an affix: like /ɪ/ of *cloudy* and /æ/ of *asymmetric*, /iː/ of *e-commerce* consists of only one sound. By contrast, the combining form *-tainment* /ˈteɪmm(ə)nt/ of e.g. *docutainment* consists of eight sounds and therefore must be regarded as a bound root.

In the case of combining forms like *-babble* of *designer-babble*, *-core* of *sadcore*, and *-erati* of *geekerati*, etc. which cannot be analyzed as either roots or affixes only because of the length of their signifiers, the preference must be given to the affix solution. That is, all these combining forms can be regarded as affixes unless one can show that these forms are capable of occurring in isolation.

Note that the existence of the noun *babble*, which, according to the OED, can express the meanings:

- 'inarticulate or imperfect speech, such as that of infants; prattle'
- 'idle, foolish, or unseasonable talk; prating'
- 'confused murmur, as of a stream'

does not suffice to regard the combining form *-babble* of e.g. *designer-babble* as a root. There is no doubt that the latter is semantically related to the former: according to the OED, the combining form *-babble* actually goes back to the noun *babble*. However, *-babble* of *designer-babble* does not refer to an inarticulate or improper speech, an idle or a foolish talk, or a confused murmur. *-Babble* of *designer-babble* refers to a particular jargon used by designers. Accordingly, given the semantic non-identity between the noun *babble* and the combining form *-babble,* the existence of the former cannot serve as a justification for the root analysis of the latter.

A different case is represented by the combining form *must-* of e.g. *a must-read, a must-see, a must-have*, etc., which 'denote things that are essential, obligatory, or highly recommended' (OED). The combining form *must-* of *must-read* seems to be semantically identical with the verb *must* of e.g. *You must read this book*. That is, a must-read is a book, an article, etc. which other people think you must read. Accordingly, *must* of *must-read* is not an affix but a root. (Given the non-bound character of *must* in words like *must-read, must-see, must-have*, etc., it is doubtful that this morph can be regarded as a combining form.)

Returning to the question raised in the title of this part of Section 2.5, we can say that so-called combining forms do not constitute a distinct morpheme type in English. Combining forms are bound morphs that are semantically different from prototypical affixes in that they express less abstract meanings. However, they typically possess all other properties characteristic of affixes: they are

The internal structure of English words 55

bound morphs, which do not occur in isolation, and their signifiers are not much longer than those of prototypical affixes. Accordingly, in the overwhelming majority of cases combining roots can be analyzed as affixes.

2.5.3 One signifier → both a root and an affix

As illustrated by the combining form *-babble* of e.g. *designer-babble* and the noun *babble*, one and the same signifier can occur both as an affix and a root. A similar example involving a prototypical affix is *out-* of verbs like *outdo*, *outnumber*, *outsell*, etc. Despite the fact that the signifier *out* can occur in isolation, *out-* of these verbs cannot be regarded as a root. Let us compare the meanings of *out* in (52) and in the previously mentioned verbs *outdo*, *outnumber*, and *outsell*.

(52)   *You will be the first student who will NOT be in my violin class. Out!*
       (COCA)

The *out* of (52) is an adverb whose meaning is described by MWO as 'in a direction away from the inside or center' / 'outside'. By contrast, the *out-* of *outdo*, *outnumber*, and *outsell* carries the meaning 'in a manner that exceeds or surpasses and sometimes overpowers or defeats' (MWO): i.e. *outnumber* means 'to exceed in number'; *outsell* means 'to exceed in selling'; *outdo* means 'to surpass in doing or performing'.

It is important to observe that the latter meaning is expressed by *out-* only when it is followed by verbs like *do*, *number*, and *sell*: it never carries this meaning when used in isolation, i.e. in contexts like that of (52). Accordingly, this is clearly a bound affixal meaning, so that *out-* of *outdo*, *outnumber*, and *outsell* must be regarded as an affix, not a root. (Note that *out-* of *outnumber* cannot be regarded as a morfoid. In the case of *outnumber* and verbs alike, there is no idiomaticity: in all these verbs *out-* carries the non-idiomatic meaning 'in a manner that exceeds or surpasses', which it does not express when used in isolation.)

Now, let us also consider the status of *out* in the verb *outsource* of (53).

(53)   *The company outsources many of its jobs to less developed countries*
       (MWO)

As defined by the OED, *outsource* means 'to obtain (goods, a service, etc.) by contract from an outside source; to contract (work) out'. (The overall meaning of *outsource* thus contains the meanings of both of its components *out* and *source* plus the idiomatic meaning 'to obtain from', which is not inherent in either *out* or *source.*) Given this definition, we can conclude that in contrast to *out-* of e.g. *outnumber*, *out* of *outsource* carries the meaning 'outside', which it also carries when used in isolation. Accordingly, in contrast to *out-* of *outdo, outnumber*, and *outsell*, *out* of *outsource* is not an affix but a root.

In summary, the signifier *out* can occur both as a root and an affix. When used in isolation and in words like *outsource* where it carries the meaning 'outside', *out* must be regarded as a root. In contrast, when occurring in verbs like *outdo, outnumber*, and *outsell* where it carries the meaning 'in a manner that exceeds or surpasses', *out-* must be regarded as an affix.

2.5.4 Typology of affixes

With respect to their position in relation to the root, English affixes are traditionally classified into the following four categories:

1. **prefixes**
2. **suffixes**
3. **interfixes**
4. **infixes**

Prefixes are affixes that precede the root: e.g. *un-* of *untrue*, *out-* of *outnumber*, etc. Suffixes are affixes that follow the root: e.g. *-er* of *teacher*, *-y* of *cloudy*, etc. Interfixes are meaningless affixes that serve as interlinking elements between either two roots or one root and one meaningful suffix: e.g. *-o-* of *morphology*. Infixes are suffixes that 'break' the root. For example, the slang word *edumacation* 'an education that someone received at a crappy school, or a lack of education all together' (Urban Dictionary) contains the infix *-ma-* 'poor quality', which is inserted into the root *educate*.

With regard to their function, affixes are traditionally classified into **inflectional** and **derivational affixes**. The most important differences are as follows.

- Inflectional affixes express obligatory grammatical meanings; derivational affixes express optional lexical meanings.

- Inflectional affixes are more productive than derivational affixes.

- Forms differing only with regard to the presence / absence of some inflectional affix always express the same lexical meaning; forms differing only with regard to the presence / absence of some derivational affix always express different lexical meanings.

- Forms differing only with regard to the presence / absence of some inflectional affix are always members of the same word class; forms differing only with regard to the presence / absence of some derivational affix are very often members of different word classes.

- Derivational affixes expressing temporal concepts can be iterated; inflectional affixes expressing temporal concepts cannot be iterated.

As in 2.5.1, let us elaborate on each of these differences. As was already mentioned in 2.1, lexical meanings are optional meanings which are expressed only when language users specifically want them to be expressed. Grammatical meanings, by contrast, are obligatory meanings which cannot be unexpressed. To illustrate this difference, let us again compare the past tense suffix *-ed* of e.g. *I worked* and the past time prefix *ex-* of e.g. *ex-president*. As we established in 2.3, these affixes express almost identical meanings and, accordingly, can be regarded as allomorphs of the same morpheme, unless one can show that *ex-* has an additional negative meaning, which is not inherent in *-ed*. However, while the past tense meaning inherent in *-ed* is an obligatory grammatical meaning, the past time meaning inherent in *ex-* is an optional lexical meaning. Thus in stark contrast to verbs, which in English are almost always marked with regard to tense, English nouns are essentially tenseless. That is, while the absence of *-ed* typically locates the verb in question (which forms the past tense with *-ed*) in the present (e.g. *talk, walk, work*), the absence of *ex-* does not necessarily mean that the noun in question denotes a non-past characteristic. For example, the noun *president* can be used in connection with a person who is no longer president, even when it is not preceded by the past time marker *ex-*. E.g. (54).

(54)   *In 2005, President Clinton established CGI to turn ideas into action and to help our world move beyond the current state of globalization [...]* (http://tinyurl.com/26jko5z)

Note also that *ex-* is a non-obligatory past time marker for nouns: instead of saying *ex-president*, we can say *former president*. By contrast, *-ed* is an obligatory past tense marker for verbs like *talk, walk, work*: no other morph can be used instead of *-ed* in combination with these verbs. Given these differences, we are justified in concluding that *-ed* is an inflectional suffix and *ex-* is a derivational prefix.

In addition to the obligatoriness–optionality distinction, inflectional affixes are usually more **productive** than derivational affixes. For example, the past tense marker *-ed* can combine with almost any English verb: e.g. *talked, worked, destroyed, deduced, turned*, etc. In contrast, the past time marker *ex-* combines with a relatively small number of nouns. According to the OED, it occurs with nouns denoting 'titles of office or dignity' (e.g. *ex-president*) or 'designating persons with respect to their calling, station, character, etc.' (e.g. *ex-boyfriend*). However, it does not seem to occur with object nouns. For instance, looking at what has remained of his or her house after an earthquake or a tornado, a speaker of English will most likely not say *This is my ex-house* meaning 'this used to be my house in the past'. Likewise, the formations *ex-chair, ex-room, ex-table*, etc. do not seem to be possible in similar contexts.

English inflectional affixes never change the lexical meaning of their **input signifiers** (i.e. signifiers to which they are attached). That is, both the singular *book* and the plural *books* have the same lexical meaning 'book'; both the present tense *work* and the past tense *worked* have the same lexical meaning 'work'; both the non-progressive *read* and the progressive *am reading* have the same lexical meaning 'read'; etc. By contrast, the addition of a derivational affix always results in a change of lexical meaning. For example, while *teach* means 'to teach', *teacher* means 'performer of the action of teaching'; while *cloud* means 'cloud', *cloudy* means 'full of clouds'; while *number* means 'number', *outnumber* means 'to exceed in number'; etc.

English inflectional affixes never change the word class of their input signifiers. For example, *book* and *books* are both nouns; *work* and *worked* are both verbs; *pretty* and *prettier* are both adjectives; etc. By contrast, derivational affixes often change the word class: e.g. *teach* is a verb, but *teacher* is a noun; *cloud* is a *noun*, but *cloudy* is an adjective; *number* is a noun, but *outnumber* is a verb, etc.

Derivational affixes expressing temporal concepts can be iterated. For instance, in order to express the meaning 'former boyfriend prior to another former boyfriend', we can say *ex ex-boyfriend*. By contrast, we cannot say *workeded in order to express a similar **anterior meaning** 'had worked' (i.e. an action that took place before some other action in the past).

Very often it is also argued that inflectional affixes are **semantically more regular** than derivational affixes. This means that anisomorphic complex words, whose signifieds are not (or not entirely) representable in terms of their components' signifieds, can be found only among derivational formations like *stealer* and *writer*, but not among inflected forms like *boys, worked, prettier*, etc. Indeed, it seems that the obligatoriness of grammatical meanings must a priori disallow their idiomaticity. That is, if the meaning 'the past tense' is an obligatory grammatical meaning, then any verb that contains the past tense suffix *-ed* must express the past tense meaning: namely, that the action expressed by the root of the verb took place before the moment of utterance. (By contrast, the optionality of lexical meanings inherent in derivational affixes must sanction their idiomaticity.) However, as we will see below, this is not always the case. Consider, for example, the meanings inherent in the underlined inflectional affixes occurring in (55) and (56).

The internal structure of English words 59

(55)   CONAN: *All right, Don, thanks very much for the call. And Myron, we just have a couple of minutes left with you. I want<u>ed</u> to ask you to tell the story of your father taking you to see the Brooklyn Dodgers in 1947.*
Mr-UHLBERG: *I'll never forget that. That was April 15th, 1947. Jackie Robinson, the first African-American player to play in the white major leagues [...]* (COCA)

(56)   *I don't want to blame Christopher for all the things I can't do anymore. He is always try<u>ing</u> to interest me in inactive activities -- reading, collage, wine-tasting* (COCA)

I wanted to ask you [...] of (55) is grammatically in the past tense: cf. *I wanted to ask you* and *I want to ask you*. However, the meaning expressed by the former in (55) is clearly 'the present tense': *I wanted to ask you* means 'I want to ask you now'. If this were not the case, Mr. Uhlberg would not start telling the story of his father taking him to see the Brooklyn Dodgers in 1947.

He is always trying to interest me in inactive activities [...] of (56) is grammatically in the progressive aspect: cf. *He is always trying to interest me in inactive activities* and *He always tries to interest me in inactive activities*. However, in (56) the former does not refer to an **ongoing action** coinciding with the moment of utterance (e.g. *I am now reading a book*) but expresses a **habitual situation** recurring on a more or less regular basis. That is, *He is always trying to interest me in inactive activities* means that 'he tries to interest me in inactive activities whenever he has the opportunity to do so'.

These examples illustrate that idiomaticity is characteristic not only of complex words containing derivational affixes like *-er* of *stealer* and *worker* but also of complex forms containing inflectional affixes like *-ed* of *wanted* and *-ing* of *is trying*: the former can be used in connection with non-past events and the latter can refer to non-progressive events.

2.5.5 Typology of roots

Like affixes, roots can also be classified into those which express optional lexical meanings and those which express obligatory grammatical meanings. The latter are known as **analytic forms**. Consider, for instance, the comparative degree of the adjective *beautiful*: *more beautiful*. In contrast to e.g. *prettier*, the grammatical meaning 'the comparative degree' is expressed here not by the inflectional affix *-er* but by the free form *more*, which is capable of occurring in isolation, thus qualifying as a root.

A well-known peculiarity of the English language is that a number of grammatical meanings are expressed both inflectionally and analytically. Recall the discontinuous progressive morph *am...ing* of *I am reading*, which we discussed in 2.1. This morph represents a combination of the analytic form *am* and the inflectional affix *-ing*. Similarly, the discontinuous passive morph *was...ed* of e.g. *was replaced by* is a combination of the analytic form *was* and the inflectional affix *-ed*.

2.6 Exercises

1. Make sure you can explain each of the key terms printed in boldface (ideally, using your own examples).

2. Which of the following statements are true?

a) The signifier of a morpheme is called the morph.
b) One and the same signifier is never associated with more than one signified.
c) The signifier is the most important component of a linguistic sign.
d) Allomorphs of the same morpheme can be in contrastive distribution.
e) A morfoid is an opaque component of an idiomatic word.
f) A portmanteau morph is a morph that cumulatively expresses more than one meaning.
g) Semi-idiomatic words can be segmented either into one normal morph and one morfoid or into one normal morph and one submorph.
h) Roots are typically shorter than affixes.
i) Combining forms are always free morphs.
j) Inflectional affixes express optional lexical meanings.

3. Analyze the distribution of the following morphs. Explain your analysis.

a) *table* and *sample*
b) /d/ of /steɪd/ in e.g. *he stayed* and /ɪd/ of /æk'sɛptɪd/ in e.g. *he accepted*
c) *cat* and *mat*
d) *in-* of *intolerable* and *ir-* of *irresponsible*
e) *in-* of *inability* and *non-* of *non-ability*
f) /iˈkɑnəmi/ and /əˈkɑnəmi/ of *economy*
g) /breɪv/ of *brave* and /ˈbrɛv/ of *brevity*
h) *-er* of *teacher* and *-eer* of *auctioneer*
i) *-s* of *boys* and *-ren* of *children*
j) *-ed* of *he has worked* and *-en* of *he has taken*

4. Analyze the morphemic structure of the following words:

a) *greenhouse*

b) *table*
c) *barman*
d) *foretell*
e) *within*
f) *permit*
g) *great*
h) *after-party*
i) *redo*
j) *birdbrain*

5. Name all possible characteristics of the following units:

a) *-ist* of *biologist*
b) *has* of *he has worked hard*
c) *mat*
d) *-en* of *taken*
e) *-less* of *careless*
f) *-en* of *oxen*
g) *-gress* of *progress*
h) *fore-* of *foresee*
i) *red* of *redneck*
j) *cat*

2.7 Further reading

My understanding of how words must be segmented into morphemes has been largely influenced by Plungian (2000: Ch. 2, part 1) and Mel'čuk (2001). Unfortunately, I can recommend these works only to those students of English morphology who can understand Russian. Those who can understand French are referred to Mel'čuk (1997), which is the original French version of Mel'čuk (2001). A concise English introduction to the main concepts of Mel'čukian morphology such as e.g. the concept of a mega-morph can be found in Mel'čuk (2006: Ch. 7); see also an earlier work Mel'čuk (1982).

Classic works dealing with the distribution of morphs are Harris (1942), Hockett (1947), and Nida (1948). If you wish to refresh your knowledge of phonetic processes like place-of-articulation assimilation, see Cruttenden (2008: especially part III). For differences between phonological and lexical conditioning of allomorphs occurring in complementary distribution, see Haspelmath and Sims (2010: 25-26). For an introduction to various semantic relations including absolute synonymy, see Cruse (2004: especially Ch. 8). A classic article dealing with the differences between polysemy and homonymy is Jakobson (1972).

Section 2.5.1, which presented the differences between roots and affixes, was largely based on Mel'čuk (2001: 69-79). Since the book is in Russian, I recommend a similar discussion in Haspelmath and Sims (2010: 19-22). A similar analysis of forms that exist both as roots and affixes can be found in Plag (2003: Sec. 4.1). For a discussion of the differences between inflectional and derivational affixes, see Haspelmath and Sims (2010: Ch. 5), Bauer (2003), Plungian (2000: Ch. 1, part 2), Jakobson (1959).

Finally, it must be noted that not all morphologists accept the views that a morpheme is a meaning-carrying unit and that words must be segmented into morphemes. If you want to become acquainted with these views, see Aronoff (1976) and Anderson (1992).

3 Analyzing English lexemes

In this chapter we will be concerned with the formal and semantic structure of English lexemes. Section 3.1 provides a definition of a lexeme. Section 3.2 deals with the distribution of English lexemes as well as their typology from both a formal and a semantic point of view. As regards the distribution of lexemes, the section will argue that like morphs that realize two different morphemes, lexes that realize two different lexemes occur in contrastive distribution, while allolexes of the same lexeme, like allomorphs of the same morpheme, can either be in complementary distribution or in free variation. Section 3.3 introduces the concept of a vocable and discusses all possible semantic relations that can hold between two or more lexemes which form the same vocable. Finally, Section 3.4 briefly touches on the concept of a lexeme family.

3.1 What is a lexeme?

A lexeme can be defined as a conventionalized association between any free form and some particular lexical meaning. For example, the conventionalized association between the signifier *cat* and the lexical meaning 'cat' forms the lexeme CAT; the conventionalized association between the signifier *untrue* and the lexical meaning 'not faithful' forms the lexeme UNTRUE; the conventionalized association between the signifier *waithood* and the lexical meaning 'a particular waiting stage in the life of a college graduate' forms the lexeme WAITHOOD; etc.
   Consider also the idiomatic VP *kicked the bucket* of (57) and the idiomatic clause (58).

(57)   *There's the guy who <u>kicked the bucket</u> after a shot* (COCA)
(58)   *A bird in the hand is worth two in the bush* (COCA)

The VP *kicked the bucket* is a fully-idiomatic phrase whose meaning contains only the grammatical past tense meaning of the inflectional affix *-ed* but not the literal lexical meanings of the components *kick*, *the*, and *bucket*: *kicked the bucket* does not mean 'kicked some particular bucket' but 'died'. Similarly, (58) is a fully-idiomatic clause which does not mean 'a bird in some particular hand is worth two in some particular bush' but 'it's better to have a small real advantage than the possibility of a greater one' (The Phrase Finder). Idiomatic clauses and sentences like (58) are traditionally called **proverbs**. According to Dobrovol'skij and Piirainen (2005: 51), the defining characteristic of a proverb is that its meaning contains the semantic component [recommendation (that is supposed to be of a universal character) how to behave in a particular situation], i.e. e.g. (58) recommends that we always choose a small real advantage rather than the possibility of a greater one.

Both the fully-idiomatic VP *kick the bucket* and the fully-idiomatic proverb *A bird in the hand is worth two in the bush* are similar to the words *cat*, *untrue*, and *waithood* in that like the latter, the former are also stored in the mental lexica of many English speakers. That is, those English speakers who know that the underlined VP of (57) and Sentence (58) do not mean what their overt components literally stand for can be said to have conventionalized associations (stored in their mental lexica) between the signifiers *kick the bucket* / *A bird in the hand is worth two in the bush* and the signifieds 'to die' / 'it's better to have a small real advantage than the possibility of a greater one'. Accordingly, it makes sense to regard idiomatic phrases and sentences as lexemes as well. (The term 'lexeme' should thus not be equated with the term 'word', even though the overwhelming majority of lexemes are words.)

3.2 The structure of a lexeme

Like other signs, lexemes have four structural components:

1. **the signifier**
2. **the signified**
3. **the syntactics**
4. **the sociolinguistics**

For example, the lexeme CAT consists of the signifier *cat*, the signified 'cat', the syntactics 'the signifier *cat* is a noun and thus can head NPs', and the sociolinguistics '*cat* is a stylistically neutral signifier and thus can occur in all possible contexts'. Similarly, the fully-idiomatic lexeme KICK THE BUCKET consists of the signifier *kick the bucket*, the signified 'to die', the syntactics '*kick the bucket* is a VP and thus can fill the predicate position', and the sociolinguistics '*kick the bucket* is a very informal way of expressing the concept of dying'.

3.2.1 The lex of a lexeme

Just as the signifier of a morpheme is referred to as its morph, the signifier of a lexeme can be referred to as its **lex**. For example, the signifier *cat* is the lex of the lexeme CAT. The signifier *untrue* is the lex of the lexeme UNTRUE. The signifier *kick the bucket* is the lex of the lexeme KICK THE BUCKET.

As in the case of phonemes and morphemes, one and the same lexeme can be realized by more than one lex. Compare, for example, the lexes *kick the bucket* and *die*. Both of them express the meaning 'to die' and can therefore be regarded as **allolexes** of the same lexeme. Since *kick the bucket* is an informal way of expressing the meaning 'to die', whereas *die* is stylistically neutral, we can say that the allolexes *kick the bucket* and *die* occur in stylistically relevant free variation. Similarly, the lexes *fever* and *pyrexia*, which are associated with the same signified 'abnormal elevation of body temperature', but differ with regard to their sociolinguistics, can be considered allolexes of the same lexeme which occur in stylistically relevant free variation.

In addition to free variation, allolexes realizing the same lexeme can also be in complementary distribution. Recall the pair *happy–happily*, which we discussed in 2.2.3. As we established, the adjective *happy* and the adverb *happily* express the same lexical meaning but occur in different syntactic environments: there can only be *a happy life* and *His life was happy*, but not **a happily life* and **His life was happily*. Accordingly, *happy* and *happily* can be regarded as allolexes of the same lexeme which occur in complementary distribution.

A special type of complementary distribution of allolexes realizing the same lexeme is represented by the distribution of **wordforms** of the same lexeme. Wordforms are allolexes that express different grammatical meanings. For example, *book* and *books*, *pretty* and *prettier*, *worked* of *he worked* and *has worked* of *he has worked*, etc. Wordforms which realize the same lexeme typically occur in identical syntactic environments: e.g. *this book* and *these books*; *She is pretty* and *She is prettier than his ex-girlfriend*; *He worked hard* and *He has worked hard*; etc. In other words, both the singular and the plural wordform of a noun can head NPs; both the positive and the comparative wordform of an adjective can function as complement of the auxiliary *be*; both the simple past and the present perfect wordform of a verb can function as predicator in a predicate VP; etc. However, despite these facts, this textbook suggests that the distribution of wordforms of the same lexeme should be regarded as an instance of complementary distribution rather than of free variation. Thus, there can only be *these schools* and *She is prettier than his ex-girlfriend*, but not **this schools* and **She is pretty than his ex-girlfriend*. Similarly, the simple past and the present perfect wordform of a verb are usually not interchangeable: both denote events that took place in the past, but the present perfect wordform is used to refer to a past event that is somehow relevant in the present: *He has worked hard (in the past) and, because of this, he is now the CEO of Microsoft*. Accordingly, wordforms of the same lexeme that have different grammatical meanings are usually not used in entirely identical environments and can therefore be said to occur in complementary distribution.

3.2.2 The typology of lexes

As we learned in Section 3.1, the lex of a lexeme can be represented by a word, a phrase, and a clause / a sentence. As we already know, words that represent lexes of lexemes can be classified into simple and complex words. The latter are words like *untrue, boyfriend, understand, receive, history,* etc. which can be segmented into at least two morphs or quasi-linguistic units. The former are words like *true, boy, friend, stand,* etc. which consist of no more than one morph.

Complex words can be further classified into **compounds** and **affixed or derived words**. Compounds are complex words like *boyfriend* that contain at least two roots. Affixed or derived words are complex words like *untrue* that contain at least one root and one derivational affix.

3.2.3 The signified of a lexeme

From a semantic point of view, lexemes realized by complex lexes can be classified into **isomorphic** and **anisomorphic lexemes**. Isomorphic lexemes are realized by complex lexes like *untrue* and *unable* whose signifieds contain only their components' signifieds. Anisomorphic lexemes are realized by complex lexes like *understand, blackboard, football,* etc. whose signifieds are not (or not entirely) representable in terms of their components' signifieds.

According to Mel'čuk (2001: 447-460), anisomorphic lexemes fall into three categories:

1. **full-idioms**
2. **semi-idioms**
3. **quasi-idioms**

A full-idiom is a lexeme realized by a complex lex like *understand* whose signified contains none of its components' signifieds. For example, the lexeme UNDERSTAND is a full-idiom because the signified 'to understand' does not contain the signifieds 'under' and 'to stand', inherent in the submorphs *under-* and *stand* when they occur in other environments.

A semi-idiom (or a **partial idiom**) is a lexeme realized by a complex lex like *blackboard* whose signified contains at least one of its components' signifieds. For example, the lexeme BLACKBOARD is a semi-idiom because the signified 'blackboard' contains only the signified 'board', inherent in the morph *board*, but not the signified 'black', inherent in the morfoid *black*.

Finally, a quasi-idiom is a lexeme realized by a complex lex like *football* whose signified contains not only its overt components' signifieds but also some

Analyzing English lexemes

additional unpredictable signified. For example, the lexeme FOOTBALL is a quasi-idiom because the signified 'football' contains not only the signifieds 'foot' and 'ball', inherent in the morphs *foot* and *ball*, but also the idiomatic signified 'a particular sport different from handball, basketball, volleyball, etc.'.

3.2.4 Three-component anisomorphic lexemes

All examples of anisomorphic lexemes that have been dealt with so far are lexemes realized by two-component complex lexes: UNDERSTAND and BOYFRIEND are fully-idiomatic lexemes realized by two-component complex lexes *understand* and *boyfriend*; BLACKBOARD and TWILIGHT are semi-idiomatic lexemes realized by two-component complex lexes *blackboard* and *twilight*; FOOTBALL, STEALER, WRITER, WAITHOOD are quasi-idiomatic lexemes realized by two-component complex lexes *football, stealer, writer, waithood*. In the following we will become acquainted with anisomorphic lexemes which are realized by complex lexes consisting of more than two components.

Compare, for example, the signifieds associated with the lexes *forget-me-not, mother-in-law*, and *nationalism*. The lexeme FORGET-ME-NOT is an obvious full-idiom whose signified '[...] a plant which flourishes in damp or wet soil, having bright blue flowers with a yellow eye' (OED) does not contain the signifieds 'to forget', 'me', 'not', inherent in the morfoids *forget, me, not* when they occur in other environments. The quasi-linguistic units *forget, me, not* can be regarded as morfoids rather than submorphs because the literal meanings inherent in these components seem to partially motivate the idiomatic meaning 'forget-me-not'. According to the OED,

> in the 15th century the flower was supposed to have the virtue of ensuring that those wearing it should never be forgotten by their lovers. (OED)

Those speakers of English who do not remember this **true etymology** of *forget-me-not* can easily invent their own popular etymologies establishing a connection between the idiomatic meaning 'forget-me-not' and the literal meanings inherent in the components *forget, me, not*. For example, the name *forget-me-not* can be attributed to "the nauseous taste that it leaves in the mouth" (OED).

The lexeme MOTHER-IN-LAW is a semi-idiom whose signified 'the mother of one's spouse' (OED) contains the signified of the morph *mother* but not of the morfoids *in* and *law*. As in the case of *forget-me-not*, the literal meanings of the quasi-linguistic units *in* and *law* seem to partially motivate the idiomatic meaning 'the mother of one's spouse': since sons-in-law are married to their mothers-in-law's daughters, there seems to exist a legal relation holding between a husband and his wife's mother. Accordingly, *in* and *law* are morfoids, not submorphs.

Finally, the lexeme NATIONALISM is a quasi-idiom whose signified does not only contain the meanings of the root *nation* and the suffixes *-al* and *-ism*. If *nationalism* were an isomorphic word, it would mean something like 'a national ideology or an ideology relating to a particular nation'. Indeed, the suffix *-ism* often denotes ideologies (e.g. *socialism, capitalism*) and the suffix *-al* carries the meaning 'of, relating to, or characterized by' (MWO); e.g. *cultural, suicidal*. However, nationalism is of course much more than just an ideology relating to a particular nation. As defined by MWO, nationalism is

> a sense of national consciousness exalting one nation above all others and placing primary emphasis on promotion of its culture and interests as opposed to those of other nations or supranational groups. (MWO)

The signified 'nationalism' thus contains the idiomatic meaning 'belief in the supremacy of one's own nation and its national interests', which is not inherent in either the root *nation* or the suffixes *-al* and *-ism*.

3.2.5 Anisomorphic lexemes realized by phrases and sentences

First of all, it must be noted that isomorphic phrases and sentences, whose signifieds contain only their components' signifieds – e.g. the NP *a black cat* meaning 'some black cat' and the clause *I saw a black cat* meaning 'I saw some black cat' – are not stored in our mental lexica and thus cannot be regarded as lexes realizing lexemes. The reason for the former is the **finiteness** of our mental lexica: it is extremely uneconomical to store phrases and sentences whose signifieds can be easily arrived at by adding their components' signifieds: e.g. 'a black cat' = 'indefiniteness' + 'black' + 'cat' or 'I saw a black cat' = 'I' + 'saw' + 'indefiniteness' + 'black' + 'cat'.

By contrast, idiomatic phrases and (at least some) idiomatic sentences are usually stored in our mental lexica. This means that, for example, a speaker of English who knows that the VP *kick the bucket* means 'to die' and the proverb *A bird in the hand is worth two in the bush* means 'it's better to have a small real advantage than the possibility of a greater one' has a conventionalized association between these signifiers and these signifieds stored in his or her mental lexicon. Accordingly, as we concluded in Section 3.1, both the fully-idiomatic VP *kick the bucket* and the fully-idiomatic proverb *A bird in the hand is worth two in the bush* must be regarded as lexemes.

In addition to fully-idiomatic phrasal and sentential lexemes like those discussed above, anisomorphic lexemes realized by phrasal and sentential lexes can also have semi- and quasi-idiomatic signifieds. Compare, for example, the meanings of the underlined VPs in (59) and (60).

(59) *Answer the door,'* she said and gazed back out at the grotto (COCA)
(60) Back in Greensboro, Duke married and *started a family* (COCA)

The VP *answer the door* of (59) is a semi-idiom whose signified contains the signifieds of the components *the* and *door*, but not of the component *answer*: *answer the door* means '[after hearing the doorbell or a knock] to go to the door to see who is there' (The Free Dictionary). In other words, answering the door involves going to a particular door (the one that was knocked at). Accordingly, both the signifieds 'door' and 'definiteness' can be considered part of the signified 'to answer the door'. By contrast, answering the door may not necessarily involve answering to the person who knocked at a particular door or rang its doorbell: opening the door without literally answering to that person would also qualify as an act of answering the door. Accordingly, the meaning 'to answer' is not part of the meaning 'to answer the door'.

The VP *started a family* of (60) is a quasi-idiom whose signified contains the signifieds of its overt components *start, -ed, a,* and *family* – that is, the signifieds 'to start', 'the past tense', 'indefiniteness', and 'family' – but in addition to these signifieds, it also contains the idiomatic signified 'to conceive the first child with one's spouse (thereby starting to have a real family)' (Mel'čuk 1995: 184). In other words, what Duke did back in Greensboro was marry someone and then conceive his first child with that very person whom he married. In this way, Duke started to have a real family, i.e. a family that includes a father, a mother, and at least one child. Marrying someone without giving birth to at least one child does not suffice to start a family.

Finally, compare the semantic structure of the idiomatic pick-up lines (61) and (62), taken from http://linesthataregood.com/.

(61) Didn't I see you on the cover of *Vogue*?
(62) Would you like to have morning coffee with me?

A pick-up line is defined by the Urban Dictionary as "a conversation opener with the intent of engaging an unfamiliar person for sex, romance, or dating". Notice that pick-up lines are typically represented by idiomatic clauses and sentences like (61) and (62). Indeed, (61) is a sentential semi-idiom whose signified contains the meanings of the components *I, you, on, the, cover, of, Vogue* but not the meanings of the components *did, not,* and *see*. That is, the speaker of (61) does not really request the addressee to confirm that the former did indeed see the latter on the cover of Vogue but simply expresses a compliment: 'You are so beautiful. I think you deserve to be on the cover of Vogue'. By contrast, (62) is not a semi- but a quasi-idiom that does not mean 'Would you like to have morning coffee with me?' but 'Would you like to go to bed with me and afterwards have morning coffee with me?'. The signified of (62) thus contains the additional idiomatic meaning 'invitation to go to bed with the speaker of (62)', which is inherent in none of (62)'s overt components.

To conclude, those speakers of English who actively use (61) and (62) with the intent of initiating a sexual relationship with an unfamiliar person can be said to have conventionalized correspondences between (61) and (62) and the just named semi- / quasi-idiomatic signifieds (which can be expressed by them) stored in their mental lexica. Accordingly, sentences (61) and (62) can be regarded as sentential lexes realizing corresponding anisomorphic lexemes.

3.2.6 How to distinguish between full-, semi-, and quasi-idioms?

Given what we have learned in the previous parts of this section, this may seem a superfluous question. As we said in 3.2.3, a full-idiom is an isomorphic lexeme realized by a lex whose signified does not contain either of its components' signifieds. A semi-idiom is an anisomorphic lexeme realized by a lex whose signified contains one of its components' signifieds. Finally, a quasi-idiom is an anisomorphic lexeme realized by a lex whose signified contains not only its overt components' signifieds but also some additional, unpredictable signifieds. These definitions can be used as guidelines for deciding whether the lexeme under analysis has a full-, semi-, or quasi-idiomatic signified.

But let us again consider the fully-idiomatic lexeme BOYFRIEND. As was pointed out in 1.2.2, the meaning 'friend', inherent in the morfoid *friend*, partially motivates the idiomatic meaning 'boyfriend': boyfriends are usually perceived (by their girlfriends) as friends, i.e. as people whom they know well and regard with affection and trust. Given this fact, why can we not analyze BOYFRIEND as a semi-idiomatic lexeme whose signified contains the signified of the component *friend* but not of the component *boy*? More generally, how can we distinguish between components' signifieds which are part of lexemes' idiomatic signifieds and components' signifieds which only motivate lexemes' idiomatic signifieds, without, however, being part of those signifieds?

To answer this question, let us recall what we said about blackboards in 1.2.2. Blackboards typically have dark surfaces (and are in this respect different from whiteboards), but they are not always black: green blackboards, for instance, do occur as well. In other words, the lex *blackboard* can be used to refer to a non-black blackboard. This fact justifies the treatment of the lexeme BLACKBOARD as a semi-idiom whose signified contains the signified of the component *board* but not of the component *black*. Hence of these two components, only *board* can be regarded as a morph realizing a morpheme, whereas *black* is a morfoid whose signified only motivates the idiomatic signified 'blackboard', without, at the same time, being part of that idiomatic signified.

In a similar way, we can prove that neither the signified 'boy', inherent in the morfoid *boy*, nor the signified 'friend', inherent in the morfoid *friend*, are part of the idiomatic signified 'boyfriend'. As regards the signified 'boy', consider the use of *boyfriend* in (63).

(63)   *I am 23 and my boyfriend is 40?* (http://tinyurl.com/5tc222w)

As (63) demonstrates, the lex *boyfriend* can be used to refer to a 40-year-old man, who obviously no longer qualifies as a young male child (which the signifier *boy* literally means). As stated earlier, a boyfriend is a male of almost any age over puberty. Accordingly, we are justified in concluding that the signified 'boy', inherent in the morfoid *boy*, is not part of the idiomatic signified 'boyfriend'.

Similarly, as (64) demonstrates, there are boyfriends who are not friends.

(64)   *my boyfriend is not a friend at all. he has done me so wrong i cant even count [...] all the mistakes he has made* (http://tinyurl.com/6zxfmmj)

That is, a male person with whom a woman has romantic or sexual relations (and whom she regards as her boyfriend) may not be regarded by her as a friend in the literal meaning of this word. Nevertheless, she will use the term *boyfriend* as long as the man in question remains her regular sexual / romantic partner in a non-marital sexual relationship. This is because the signified 'friend', inherent in the morfoid *friend*, is also not part of the idiomatic signified 'boyfriend'. Hence we can conclude that the lexeme under analysis is indeed a fully-idiomatic lexeme whose signified does not contain the signifieds 'boy' and 'friend', inherent in the morfoids *boy* and *friend*.

In summary, in order to distinguish between components' signifieds which are part of lexemes' idiomatic signifieds and components' signifieds which only motivate lexemes' idiomatic signifieds, without, however, being part of those signifieds, we need to establish whether the lex under analysis can be used to refer to a person or a thing that does not possess the characteristics denoted by its components. If we discover that, for example, there are blackboards that are not black and boyfriends who are neither boys nor friends, we will be justified in claiming that the signified of the lexeme under analysis does not contain one or both of its components' signifieds.

3.3 Lexemes and vocables

As we established in 2.2.1, one and the same signifier can be associated with more than one signified. Recall that the signifier *true* can mean 'faithful' (e.g. *Her lover had been true*) and 'not false' (e.g. *Indicate whether each of the following statements is true or false*). Accordingly, we must distinguish between two different TRUE lexemes: TRUE$_1$ 'faithful' and TRUE$_2$ 'not false'. At the same time recall that these two signifieds inherent in *true* are polysemes: as we said, being not false can be explained as being faithful to the truth, so that the signified 'not false' can be regarded as a product of semantic narrowing of the signified 'faithful'. Given this fact, the two different TRUE lexemes can be united into the same **vocable** (Mel'čuk and Zholkovsky 1988: 27).

A vocable is a set of polysemous lexemes. For example, the vocable TRUE is the set of polysemous lexemes like TRUE$_1$ 'faithful' and TRUE$_2$ 'not false'. In addition to these signifieds, the signifier *true* can also mean 'properly so called' (e.g. *true love*), 'legitimate, rightful' (e.g. *our true and lawful king*), 'narrow, strict' (e.g. *in the truest sense*), etc. (MWO). Along with TRUE$_1$ and TRUE$_2$, these polysemous lexemes can be said to constitute the vocable TRUE.

Uniting polysemous lexemes into vocables is a usual lexicographic practice. Thus a **dictionary entry** for *true* (i.e. what you find in a dictionary like the OED or MWO when looking up *true*) is typically not just one signified 'faithful' but a set of polysemous signifiers like 'faithful', 'not false', 'properly so called', 'narrow, strict', etc.

Unlike polysemous lexemes, homonymous lexemes such as, for example, CASE 'a set of circumstances or conditions' and CASE 'a box or receptacle for holding something' do not form vocables or other similar units. This is because the signifieds of homonymous lexemes do not have much in common. That is, if the signifieds 'a set of circumstances or conditions' and 'a box or receptacle for holding something' have nothing in common, what is the point of uniting these signifieds into vocables or units alike?

3.3.1 Relations between members of the same vocable

As illustrated by the lexemes TRUE$_1$ 'faithful' and TRUE$_2$ 'not false', two lexemes that form the same vocable often exhibit a quasi-idiomatic relation, i.e. the signified of one of the two lexemes can be analyzed as a quasi-idiom in relation to the signified of the other lexeme. Thus the signified of the lexeme TRUE$_2$ can be analyzed as a quasi-idiom in relation to the signified of the lexeme TRUE$_1$: while TRUE$_1$ is associated with the signified 'faithful', TRUE$_2$ is associated with the signified 'faithful to the truth'. Compare also the meanings of *had* in (65) and (66).

(65)   *I always <u>had</u> a good time when Crystal and I got together* (COCA)
(66)   *He has not <u>had</u> many women, has drifted in solitude for much of his life [...]* (COCA)

In (65), *had* carries the meaning 'to experience something' (Ganshina and Vasilevskaya 1964: 126): *to have a good time* means 'to experience a good time'. The 'experience'-sense of *have* is historically posterior to the lexeme HAVE meaning 'to possess' (which perhaps can be regarded as a more basic meaning of *have*): according to the OED, the 'experience'-*have* has been known since circa 1000, whereas the 'possess'-*have* has been documented since circa 888. However, in Present-day English the 'experience'-uses of *have* as those in (65) are not perceived as idiomatic. As an illustration, consider the meanings of *have* in (67).

(67)   *You work to <u>have</u> the American dream, to <u>have</u> a nice house, a car, two dogs and a cat* (COCA)

Some native speakers of English fail to recognize that the *have* of *have the American dream* is semantically different from the *have* of *have a nice house*. While the latter *have* means 'to possess', the former *have* means 'to experience': *to have a nice house* means 'to possess a nice house', but *to have the American dream* means 'to experience the American dream'. That this difference is, however, a hardly recognizable difference corroborates the analysis of the 'experience'-sense of *have* as one of the literal meanings of the signifier *have*. What follows from this is that one and the same vocable can have more than one literal lexeme. That is, for example, the vocable HAVE can be said to consist of at least two literal lexemes: HAVE$_1$ 'to possess' and HAVE$_2$ 'to experience'.

In contrast to the literal *haves* of (65) and (67), the *had* of (66) is a lex realizing an idiomatic HAVE lexeme. The *had* of (66) carries the quasi-idiomatic meaning 'to experience sexual relations with somebody': *He has not had many women [...]* means 'he has not experienced sexual relations with many women'. That is, the meaning of the *had* of (66) contains the literal meaning 'to experience' plus the idiomatic meaning 'sexual relations'. Accordingly, we can conclude that within the vocable HAVE, there is the idiomatic lexeme HAVE$_3$, whose signified 'to experience sexual relations' represents a quasi-idiom in relation to the signified of one of the literal lexemes HAVE$_2$ 'to experience'.

Two polysemous lexemes that form the same vocable can also exhibit a fully- or a semi-idiomatic relation. As an illustration of the former, compare the meanings of *mouse* in (68) and (69).

(68)   *He taught me probably as much as anybody I coached because it was a struggle [...] and we got along like cat and <u>mouse</u>* (COCA)

(69)   *All computers have similar attributes such as a <u>mouse</u>, CD drive, key placements, visual characteristics, and operating systems* (COCA)

In (68), *mouse* carries its literal meaning 'any of numerous small rodents [...] with pointed snout, rather small ears, elongated body, and slender tail' (MWO). By contrast, the *mouse* of (69) denotes a computer mouse, i.e. a device that 'controls movement of the cursor and selection of functions on a computer display' (MWO). It is evident that the semantic relation between the two MOUSE lexemes is different from that holding between the lexemes $TRUE_1$ 'faithful' and $TRUE_2$ 'not false' and between the lexemes $HAVE_2$ 'to experience' and $HAVE_3$ 'to experience sexual relations with somebody'. The lexeme $MOUSE_2$ 'a computer device' can be analyzed as a full-idiom in relation to the lexeme $MOUSE_1$ 'a small rodent': the signified 'a computer device' does not contain the signified 'a small rodent'.

As an illustration of a semi-idiomatic relation holding between two polysemous lexemes, compare the meanings of the verb *to better* in (70) and (71).

(70)   *They are trying to <u>better</u> the lives of working people* (MWO)
(71)   *His financial situation has <u>bettered</u>* (COCA)

While the *better* of (70) means 'to make better', the *bettered* of (71) expresses the meaning 'to become better'. Accordingly, we can argue that the two TO BETTER lexemes are in a semi-idiomatic relation to each other. Both of their signifieds contain the signified 'better' but differ with regard to the verbal signified preceding the signified 'better'. As just said, while the *better* of (70) means 'to make better', the *bettered* of (71) means 'to become better'.

3.4 Lexemes and lexeme families

Finally, we will introduce another important concept in morphology, the concept of a **lexeme family**. According to Haspelmath and Sims (2010: 17), lexeme families are formed by lexemes whose lexes have the same root but different derivational affixes. For example, all lexemes realized by the lexes *true*, *untrue*, *truth*, *truthful*, etc. (i.e. those which share the root *true*) can be said to form a lexeme family. In a similar way, we can say that a lexeme family is formed by all lexemes realized by the lexes *history*, *historic*, *historical*, *historically*, *historicalness*, *historian*, etc., i.e. lexes which all share the bound root *histor-*.

Lexeme families are similar to vocables in that both consist of semantically and formally related lexemes. However, lexemes that form a vocable always share the same signifier, whereas members of the same lexeme family have

Analyzing English lexemes

different signifiers: they share the same root but have different derivational affixes.

3.5 Exercises

1. Make sure you can explain each of the key terms printed in boldface (ideally, using your own examples).

2. Which of the following statements are true?

a) A proverb is usually an idiomatic sentence that contains a recommendation as to how one should behave in a particular situation.
b) All linguistic signs can be regarded as lexemes.
c) Allolexes of the same lexeme are always in contrastive distribution.
d) Wordforms are allolexes of the same lexeme which differ with regard to their grammatical meanings.
e) The mental lexicon of the average speaker of English is infinite.
f) A complex word consisting of two roots is an affixed word.
g) A lexeme realized by a complex lex whose signified contains only its overt components' signifieds is an isomorphic lexeme.
h) A quasi-idiom is a lexeme whose signified does not contain its components' signifieds.
i) A vocable cannot consist of more than one literal lexeme.
j) Two different lexemes whose signifiers share the same root but have different derivational affixes form a lexeme family.

3. Analyze the distribution of the following lexes.

a) *cat* and *dog*
b) *they* and *their*
c) *painful* and *painfully*
d) *United States of America* and *U.S.A.*
e) *book* and *books*
f) *true* of *Her lover had been true* and *true* of *in the truest sense*
g) *sad* and *sadly*
h) *true* of *Her lover had been true* and *faithful* of *Her lover had been faithful*
i) *laboratory* and *lab*
j) *I* and *my*

4. State whether the members of the following pairs of lexes are different wordforms of the same lexeme, lexes realizing different lexemes of the same

vocable, lexes realizing different lexemes of the same lexeme family, or lexes realizing different lexemes that do not form either lexemes or vocables.

a) *table* and *sample*
b) *part* and *partial*
c) *bank* 'financial institution' and *bank* 'the ground near the river'
d) *teacher* and *teacher's*
e) *bachelor* 'lowest university degree' and *bachelor* 'an unmarried man'
f) *response* and *irresponsibility*
g) *a wife* and *to wife*
h) *to wait* and *waithood*
i) *is doing* and *has done*
j) *professor* and *university*

5. Classify the following words into both the formal and semantic categories which were introduced in this chapter.

a) *greenhouse*
b) *table*
c) *barman*
d) *foretell*
e) *skinhead*
f) *diary*
g) *browser*
h) *after-party*
i) *redo*
j) *birdbrain*

3.6 Further reading

Terms like 'lexeme', 'mental lexicon', 'complex word', 'compound', etc. are standard morphological terms that are introduced in any textbook on morphology. See, for instance, Lieber (2010: Chs. 1 and 2), Haspelmath and Sims (2010: Ch. 2), Katamba and Stonham (2006: Chs. 2 and 3).

The semantic classification of lexemes into full-idioms, semi-idioms, and quasi-idioms that was proposed in this chapter was largely based on the article Mel'čuk (1995) and the already mentioned book Mel'čuk (2001: Ch. 9), whose French original is Mel'čuk (1997: Ch. 9).

Idiomatic VPs like *kick the bucket, answer the door, start a family*, etc. have traditionally been the focus of attention of phraseology. For an overview of the most important phraseological issues, see Burger et al. (2007), Dobrovol'skij and Piirainen (2005; 2009) and references therein.

Idiomatic sentences like the two pick-up lines discussed in 3.2.5 have traditionally been dealt with in pragmatics, where they are regarded not as semi-idioms or quasi-idioms but as **indirect speech acts**, in which 'what is said' is different from 'what is meant'. A classic article dealing with this topic is Searle (1975). For an overview of the most important pragmatics-related issues, see Horn and Ward (2004).

4 Word-formation: basic issues

In Chapter 3 we classified English lexemes into a number of semantic and formal categories. For example, we said that UNTRUE is an isomorphic lexeme whose signified contains only the signifieds of the root *true* and the affix *un-*, whereas FOOTBALL is a quasi-idiomatic lexeme whose signified contains not only the signifieds of its roots *foot* and *ball* but also the idiomatic signified 'a particular sport different from other sports'. Proceeding from this analysis, we can conclude that the lexeme UNTRUE represents a product of isomorphic affixation of the root *true* by means of the prefix *un-*, whereas the lexeme FOOTBALL came into existence via the compounding of the roots *foot* and *ball* accompanied by the quasi-idiomatization of the meanings 'foot' and 'ball'. Beginning with this chapter, we will be concerned with the nature of processes like affixation and compounding with the help of which speakers of English create new words, thereby enlarging the vocabulary of their language. The first section of this chapter proposes a classification of word-formation into lexeme-formation and lex-formation. Section 4.2 dwells on different lexeme-formation mechanisms as well as on such issues as differences between diachronic and synchronic approaches to lexeme-formation, institutionalization of newly formed lexemes, productivity of various lexeme-formation mechanisms, etc. Finally, Section 4.3 discusses lex-formation mechanisms, which give rise to additional allolexes of already existing lexemes rather than to lexes realizing new lexemes.

4.1 Lexeme-formation versus lex-formation

Word-formation (understood as a formation of a new combination of sounds capable of independent use in accordance with the isolatability criterion) can be classified into lexeme-formation and lex-formation. The former gives rise to output lexes which realize new lexemes; the latter produces allolexes of already existing lexemes.

As an illustration of this difference, let us compare the lexes *untrue* and *happily*. From a formal point of view, both can be regarded as instances of affixation: the **output lex** *untrue* is a product of affixation of the **input lex** *true* by means of the prefix *un-* and the output lex *happily* is a product of affixation of the input lex *happy* by means of the suffix *-ly*. However, while the output lex *untrue* realizes a different lexeme than the corresponding input lex *true* – the signified 'untrue' is the reverse of the signified 'true' – the output lex *happily* is an allolex of the input lex *happy*: as we argued in 2.2.3, *happy* and *happily* differ with regard to their syntactic functioning but express essentially the same meaning. Accordingly, the formation of *untrue* is an instance of lexeme-formation, whereas the formation of *happily* is an instance of lex-formation.

Apart from classifying word-formation into lexeme-formation and lex-formation, we are now also in a position to conclude that the typology of affixes which was presented in 2.5.4 was not complete. Affixes fall not only into derivational and inflectional affixes. In addition to these two traditionally recognized categories, there are also **lex-forming** or **lex-building affixes** like *-ly* of *happily*. Lex-forming affixes are similar to inflectional affixes in that both form allolexes of already existing lexemes. However, while the addition of an inflectional affix gives rise to an output allolex that has a different grammatical meaning than a corresponding input lex (cf. e.g. *book* and *books*), the addition of a lex-forming affix never changes the grammatical meaning of a corresponding input lex. For example, as Pullum and Huddleston (2002: 532) observe, both adjectives and adverbs can be marked with regard to the grammatical category DEGREES OF COMPARISON or simply GRADE. This means that there can be the grammatical contrast between both the adjectival positive and comparative wordforms *happy* and *happier* and the adverbial positive and comparative wordforms *happily* and *more happily*. Accordingly, we are justified in concluding that the output adverbial lex *happily* does not differ from its input adjectival lex *happy* with regard to the grammatical category DEGREES OF COMPARISON: both express the grammatical meaning 'positive degree of comparison'. This justifies our analysis of the suffix *-ly* as a lex-forming suffix.

4.2 Lexeme-formation

In this section, we will be concerned with the most important issues pertaining to lexeme-formation. We will begin with a classification of lexeme-building mechanisms into the following three categories:

1. **purely semantic mechanisms**
2. **purely formal mechanisms**
3. **mechanisms that involve both semantic and formal modifications**

4.2.1 Purely semantic mechanisms

Purely semantic mechanisms are those that involve only semantic modifications of the signified of an already existing input lexeme. To put it in a simpler way: while the signifier remains the same, the signified undergoes **semantic change**. For example, the lexeme HAVE$_3$ 'to experience sexual relations' (e.g. *He did not*

Word-formation: basic issues   81

have many women) can be said to have come into existence via quasi-idiomatization of the lexeme HAVE₂ 'to experience' (e.g. *He did not have a good time*). Similarly, the lexeme MOUSE₂ 'a computer device' can be said to have come into existence via full-idiomatization of the lexeme MOUSE₁ 'a small rodent'.

An instance of semantic change that has received a lot of attention in English theoretical linguistics is so-called **morphological conversion**. Recall the verb *to wife* 'to downplay a woman's career accomplishments in favor of her abilities as wife and mother', which we discussed in 1.1. Like HAVE₃ 'to experience sexual relations', TO WIFE came into existence via quasi-idiomatization of the lexeme A WIFE: the signified 'to wife' contains the signified 'a wife', inherent in the component *wife*, plus the additional idiomatic meaning 'to downplay a woman's career accomplishments in favor of her abilities as mother'. However, in contrast to HAVE₃, the output lexeme TO WIFE is realized by a lex which is a member of a different word class than the lex which realizes the input lexeme A WIFE: while *a wife* is a noun, *to wife* is a verb. Similarly, the verb *to Thomas* 'to accuse a person of sexual harassment' (Word Spy) came into existence via full-idiomatization of the proper noun *Clarence Thomas*: according to Word Spy, this converted verb "originates from the sexual harassment accusations aimed at Clarence Thomas during his Supreme Court confirmation hearings". As in the case of *to wife*, the semantic change undergone by the input lexeme THOMAS gave rise to the output lexeme TO THOMAS, whose lex is a member of a different word class than the lex of the corresponding input lexeme: while *Thomas* is a proper noun, *to Thomas* is a verb.

4.2.2 Purely formal mechanisms

Purely formal mechanisms include:

1. **lexeme manufacturing**
2. **isomorphic borrowing**
3. **isomorphic affixation**
4. **apophony**

Lexeme-manufacturing can be described as an **arbitrary formation** of a previously non-existent signifier for expressing a previously unexpressed signified. According to Ferdinand de Saussure (1973: 100-103), **arbitrariness** is one of the major characteristics of the signifier–signified relation exhibited by linguistic signs. That is, in a number of cases there is no special reason why a particular signified is expressed by a particular signifier. For example, there is no special reason why the signified 'table' is expressed in English by the signifier *table*, but not by e.g. the signifier *Tisch*, which expresses the same signified in German, or by the signifier *стол / stol*, which expresses the same signified in Russian. Arbitrary formation is thus the creation of an unmotivated signifier like *table* that does not provide any clue as to why that very signifier was chosen by its creator for expressing a particular signified. Consider, for example, the famous proprietary name *Viagra* 'the drug sildenafil citrate, given orally in the treatment of male impotence' (OED). According to the OED, *Viagra* is

> probably an arbitrary formation. The first element *vi-* is perhaps influenced by English words such as *virile* [...], *virility* [...], etc., or their etymon classical Latin *vir* [...]. According to the makers of the drug, the last two syllables (and first vowel) were not suggested by *Niagara* [...], as sometimes proposed. A phonetic similarity to Sanskrit *vy-agra* 'excited, agitated' and *vyāghra* 'tiger' has also been noted, but these words seem unlikely to have influenced the formation. (OED)

A similar example of an arbitrary formation is *Gonk* 'the proprietary name of an egg-shaped doll' (OED). While *vi-* of *Viagra* could have been influenced by *vi-* of *virile* and *virility*, *Gonk* seems to be an entirely unmotivated signifier that does not provide any clue as to why it was created as the proprietary name of an egg-shaped doll.

Isomorphic borrowing is the importation of a foreign language lexeme which is not accompanied by a semantic modification of the signified of that lexeme (which is characteristic of **anisomorphic borrowing**). For example, speakers of English borrowed the lexeme KINDERGARTEN from German without changing its signified: at the moment of borrowing (1852), the German lexeme had a semi-idiomatic signified 'a school for the instruction of young children according to a method devised by Friedrich Froebel [...]' (OED). This is the signified which the borrowed German lex *kindergarten* came to be associated with in English. Similarly, the German *Sprachgefühl*, which has been recorded in English since 1902, came to be associated with essentially the same quasi-idiomatic signified 'the intuitive feeling of a speaker for the essential character of a language; linguistic instinct' (OED), which this signifier expressed in German.

Isomorphic affixation gives rise to new lexemes like UNTRUE and UNABLE whose signifieds are fully-representable in terms of their components' signifieds. That is, 'untrue' = 'not' + 'true' and 'unable' = 'not' + 'able'. More recent instances of isomorphic affixation by means of the derivational prefix *un-* are *unsackable* 'not sackable' (OED) and *uncreolized* 'of a language or dialect: not creolized; that has not undergone creolization' (OED). According to the OED, both have been used in English since 1980.

Finally, **apophony** can be defined as any isomorphic modification of the lex of an input lexeme that cannot be regarded as an instance of isomorphic affixation. Consider, for instance, the pair *to increase–an increase*. At first glance, it may seem that AN INCREASE is a product of quasi-idiomatization of TO INCREASE: the signified of the former lexeme can be described as 'the action of increasing' (OED). However, in contrast to the members of the conversion pairs *a wife–to wife* and *Thomas–to Thomas*, the signifiers *to increase* and *an increase* are not identical. While the verb /ɪnˈkriːs/ is stressed on the last syllable, the noun /ˈɪnkriːs/ is stressed on the first syllable. Accordingly, the difference in meaning between these two lexemes can be attributed to the shifting of the stress: /ɪnˈkriːs/ → /ˈɪnkriːs/. This formal modification cannot qualify as an instance of affixation because stress is not an affix but a **non-segmental property** of words, i.e. one that cannot be represented in terms of sounds. Consequently, /ɪnˈkriːs/ → /ˈɪnkriːs/ can be regarded as an instance of apophony. A more recent example is *buildering* 'the practice of climbing up large, urban buildings [...] as a recreational activity' (OED), which differs from its input lex *bouldering* 'practice climbing on large boulders' (OED) only with regard to its first vowel: cf. /ˈbɪld(ə)rɪŋ/ and /ˈbəʊldərɪŋ/. As in the case of *to increase* → *an increase*, the vowel change [əʊ] → [ɪ] exemplified by the pair *bouldering–buildering* does not qualify as an instance of affixation: neither [əʊ] of *bouldering* nor [ɪ] of *buildering* can be regarded as either morphs or quasi-linguistic units of the lexes under analysis. Hence *bouldering* → *buildering* can also be seen as an instance of apophony.

4.2.3 Mechanisms involving formal and semantic modifications

These mechanisms include:

1. **compounding**
2. **blending**
3. **idiomatization of phrases and sentences**
4. **anisomorphic affixation**
5. **back-formation**
6. **anisomorphic borrowing**

Compounding is an anisomorphic lexeme-building mechanism that produces compound lexemes (i.e. lexemes whose lexes are segmentable into at least two roots) whose signifieds are not (or not entirely) representable in terms of their components' signifieds. That is, those compounds that are neither semi- nor full-idioms can be analyzed as quasi-idioms, whose signifieds contain some additional, unpredictable signifieds, which are not inherent in their roots' signifieds. Consider, for example, the signifieds of the compounds *piano-tuner*, *brake cable*, and *spring festival*. As argued by Haspelmath and Sims (2010: 191), these are non-idiomatic compounds whose signifieds are made up of their components' signifieds. However, *spring festival* does not mean 'any festival that has something to do with spring' but 'a festival that takes place in spring'. Given the components *spring* and *festival*, it could have acquired a meaning like 'a festival devoted to the celebration of spring that takes place in winter or any other season of the year'. The compound *spring festival* is thus an obvious quasi-idiom whose signified does not only contain the signifieds 'spring' and 'festival', inherent in the roots *spring* and *festival*, but also the idiomatic meaning 'takes place in'. Similarly, *brake cable* is a quasi-idiom whose signified does not only contain the signifieds 'brake' and 'cable', inherent in the roots *brake* and *cable*, but also the idiomatic meaning 'a component part of': a brake cable is not any cable that has something to do with a brake but a cable that is a component part of a brake. Finally, a piano-tuner is not a person who tunes some piano but a person who tunes (more than one) particular pianos: those that other people ask him or her to tune. Again, we see that the signified 'piano-tuner' is not entirely representable in terms of the signifieds 'piano', 'tune', and 'performer of some action', inherent in the roots *piano* and *tune* and the derivational suffix *-er*.

Blending is an anisomorphic lexeme-building mechanism that produces blended lexemes, i.e. lexemes whose lexes contain the shortened lexes of (at least some of) their input lexemes. For example, *Brangelina* ← B~~rad P~~itt ~~and~~ *Angelina* ~~Jolie~~, *Chermany* ← C~~hina and~~ *Germany*, etc. From a semantic point of view, the signified of a blend is, like the signified of a compound, never fully-representable in terms of its components' signifieds. Thus *Brangelina* does not mean 'Brad Pitt' + 'Angelina Jolie' but 'a celebrity supercouple consisting of Brad Pitt and Angelina Jolie' (Wikipedia). Similarly, *Chermany* does not mean 'China' + 'Germany' but 'China and Germany taken together, particularly as an economic entity or market' (Word Spy). Both the blends *Brangelina* and *Chermany* are thus quasi-idioms whose signifieds contain additional signifieds 'supercouple' / 'economic entity', which are not inherent in the signifieds of these blends' shortened components.

Idiomatization of phrases and sentences is an anisomorphic lexeme-building mechanism that produces lexemes which are realized by phrasal and sentential lexes like those discussed in 3.2.5. For example, the semi-idiomatic lexeme TO ANSWER THE DOOR is realized by the VP *answer the door*; the fully-idiomatic lexeme A BIRD IN THE HAND IS WORTH TWO IN THE BUSH is realized by the clause *A bird in the hand is worth two in the bush*; etc.

Anisomorphic affixation is the reverse of isomorphic affixation, which we discussed in the previous part of this section. Consider, for example, the lexeme UNGOOGLEABLE 'a person for whom no information appears in an Internet search engine, particularly Google' (Word Spy). In contrast to the signifieds of the isomorphic lexemes UNTRUE and UNABLE, which came into existence via affixation of the input lexes *true* and *able* by means of the prefix *un-*, the signified 'ungoogleable ' does not only contain the signifieds 'not', 'Google', and 'able', which are inherent in the components *un-*, *Google*, and *able*, but also the idiomatic signified 'information about a person that is not available on the Internet'. A similar example is *endism* 'the belief that something of significant scope and duration, particularly something negative, is coming to an end' (Word Spy). Like the signified 'ungoogleable', the signified 'endism' does not only contain the signifieds 'end' and 'belief, ideology', inherent in the root *end* and the suffix *-ism*, but also the idiomatic meaning 'something of significant scope and duration coming to'. That is, *endism* is not the belief in the end, but the belief that something of significant scope and duration will come to an end. Both *ungoogleable* and *endism* are thus quasi-idioms, whose signifieds are not entirely representable in terms of their components' signifieds.

Back-formation is the removal of a derivational affix (or a part of an input lex which is perceived as a derivational affix) from the lex of an input lexeme. For instance, the output verb *to tase* 'to use a Taser on (a person); to subdue or incapacitate using a Taser' (OED / 1991, i.e. the date of the earliest citation documented by the dictionary) came into existence via the removal of *-er* from the input noun *Taser* 'a weapon which fires barbs attached by wires to batteries, and causes temporary paralysis' (OED / 1972). Similarly, the output noun *skeeve* 'an obnoxious or contemptible person; a person regarded as disgusting, unpleasant, etc.' (OED / 1987) came into existence via the removal of *-y* from the input adjective *skeevy* 'disgusting, distasteful, or dirty; discomforting; sleazy' (OED / 1976). As regards the semantic perspective, observe that the output signifieds 'to tase' and 'skeeve' can be analyzed as quasi-idioms in relation to their input signifieds 'Taser' and 'skeevy': *to tase* means 'to use a Taser on somebody' and *skeeve* means 'a skeevy person'. Accordingly, we are justified in regarding back-formation as an anisomorphic lexeme-building mechanism.

Finally, **anisomorphic borrowing** is the importation of the signifier of a foreign language lexeme accompanied by a semantic modification of its signified. Consider, for instance, the lexeme WIKI 'a type of web page designed so that its content can be edited by anyone who accesses it, using a simplified markup language' (OED / 1995). According to the OED, the lex *wiki* was borrowed into English from Hawaiian. However, whereas the input Hawaiian lexeme is associated with the signified 'quick', the corresponding output lexeme in English came to be associated with the signified 'a type of Web page'. Accordingly, we can conclude that the borrowing of the lex *wiki* from Hawaiian into English was accompanied by the full-idiomatization of its input signified 'quick'.

4.2.4 Diachronic and synchronic perspectives

The study of lexeme-formation can be approached from two different theoretical perspectives:

1. **lexeme-formation as a diachronic history of a particular lexeme**
2. **lexeme-formation as a synchronic relation holding between two lexemes**

As an illustration of the difference between the two approaches, let us again consider the word *boyfriend*. Thanks to the OED (the most important tool for the diachronic study of lexeme-formation in English), we know that the fully-idiomatic lexeme BOYFRIEND$_2$ 'a male of almost any age over puberty with whom a woman has a non-marital sexual relationship' is not a product of anisomorphic compounding of the roots *boy* and *friend* but a product of full-idiomatization of the lexeme BOYFRIEND$_1$ 'a friend who is a boy, a boyhood friend' (OED). According to the OED, the quasi-idiomatic BOYFRIEND$_1$ has been used in English since 1822, whereas the fully-idiomatic BOYFRIEND$_2$ appeared only in 1906. This is the diachronic history of BOYFRIEND$_2$, which allows us to conclude that the lexeme under analysis is a product of a purely semantic lexeme-building mechanism, not compounding.

However, from a synchronic perspective, a (slightly) different analysis is called for. According to the OED, the original lexeme BOYFRIEND$_1$ does not occur very often in Present-day English. (In this respect, BOYFRIEND$_1$ is different from GIRLFRIEND$_1$ 'a female friend; especially a woman's close female friend' (OED), which still coexists with GIRLFRIEND$_2$ 'a female of almost any age over puberty with whom a person has a non-marital sexual relationship'.) This means that for the majority of English speakers, the vocable **BOYFRIEND** consists of only one fully-idiomatic lexeme BOYFRIEND 'a male of almost any age over puberty with whom a woman has a non-marital sexual relationship', whose lex is easily segmentable into the components *boy* and *friend* capable of occurring in isolation. Accordingly, laymen who do not know the diachronic history of this lexeme can 'falsely' assume that BOYFRIEND is a product of anisomorphic compounding of the roots *boy* and *friend*.

Another interesting case is represented by the pair *king* / *queen–royal*. Semantically, the signified 'royal' can be analyzed as a quasi-idiom in relation to the signifieds 'king' and 'queen': according to the OED, the lex *royal* is associated with the signified 'befitting or appropriate to a monarch, especially in quality, size, or ostentation'. That is, the signified 'royal' contains the signifieds 'king' / 'queen' plus the signified 'befitting to'. However, despite this obvious semantic connection, the signifiers *king* / *queen* and *royal* do not seem to have much in common. Cf. /kɪŋ/ and /ˈrɔɪəl/. The reason for this is that *king* and *queen* are lexes of Germanic origin, which have been used in English since the Old

English period, whereas *royal* was borrowed into English from Middle French around 1400 (OED). This is the diachronic history of the lexeme ROYAL, which accounts for the fact that it is realized by a lex which has virtually nothing in common with the lexes of the semantically related lexemes KING and QUEEN.

However, as in the case of BOYFRIEND₂, we can imagine that a considerable number of laymen no longer remember that *royal* is a lex of French origin which was borrowed into English in 1400. For them the output lexeme ROYAL is a product of **suppletion** of the input lexemes KING and QUEEN. Suppletion can be defined as the use of a formally unrelated signifier (i.e. one which has a different root) for expressing a related signified. As Apresjan (1974: 172) points out, Present-day English abounds in **suppletive pairs**: in addition to *king / queen–royal*, there are also the pairs *town–urban, law–legal, father–paternal, noun–nominal, tense–temporal*, etc. Suppletive pairs can also often be found among geographical terms. For example, an inhabitant of Manchester is usually called a *Mancunian* and the language of the Netherlands is usually called *Dutch* (even though the affixed lexes *Manchesterian* and *Netherlandish* do exist as well). As in the case of *king / queen–royal*, these suppletive pairs came into existence because the lexes which realize corresponding adjectival lexemes were borrowed from other languages: *urban, legal, nominal, temporal,* and *Mancunian* were borrowed from Latin, *paternal* was borrowed from French, and *Dutch* was borrowed from Middle Dutch (OED).

Finally, let us consider the fully-idiomatic verb *babysit*. From a formal point of view, this verb can be segmented into the morfoids *baby* and *sit*, whose signifieds 'baby' and 'to sit' partially motivate the signified 'to babysit', without, at the same time, being part of that signified. *Babysit* does not mean 'to sit with or near someone's baby' but 'to care for children usually during a short absence of the parents' (MWO): we can imagine a hyperactive babysitter who never sits while doing the job of a babysitter. Evidently, this person will nevertheless qualify as a babysitter as long as he or she takes care of the baby whom he or she is supposed to babysit: it does not really matter whether a babysitter sits or stands with or near someone's baby while doing babysitting. Accordingly, we can conclude that the signified 'to sit', inherent in the morfoid *sit*, is not part of the idiomatic signified 'to babysit'. Similarly, the signified 'baby', inherent in the morfoid *baby*, is also not part of the signified 'to babysit': a babysitter can care for children of all ages, not for babies only.

Thanks to the OED, we know that the verb *babysit* appeared later than the noun *babysitter*: while the latter has been used in English since 1937, the earliest citation of the former provided by the OED dates 1947. Accordingly, we can conclude that the verbal lexeme TO BABYSIT is, from a diachronic point of view, a product of back-formation of the nominal lexeme BABYSITTER: the output lex *babysit* came into existence via the removal of *-er* from the input lex *babysitter*.

But what is the synchronic relation holding between the lexemes TO BABYSIT and BABYSITTER? Do laymen who do not know that the former appeared after the latter nevertheless analyze the verbal lexeme as a back-derivative of the corresponding nominal lexeme? As argued by Marchand (1969: 394), this is indeed the case. That is, from a synchronic point of view, TO BABYSIT must also be regarded as a back-derivative of BABYSITTER because the signified 'to babysit' is more complex than the signified 'babysitter': whereas *babysitter* means 'babysitter', *to babysit* means 'to do the job of a babysitter'. A similar example analyzed by Marchand is the verbal lexeme PROOFREAD, which, like BABYSIT, qualifies as an instance of back-formation from a diachronic point of view: according to the OED, TO PROOFREAD appeared later than PROOFREADER. As Marchand argues, the former can be considered a back-derivative of the latter not only from a diachronic but also from a synchronic perspective because the signified 'to proofread' is more complex than the signified 'proofreader': whereas *proofreader* means 'proofreader', *to proofread* means 'to do the job of a proofreader'.

This textbook rejects these analyses. From a synchronic point of view, TO BABYSIT is a fully-idiomatic lexeme whose signified does not contain the signifieds of the components *baby* and *sit*. That is, speakers of Present-day English know that *to babysit* does not mean 'to sit with or near someone's baby' but 'to take care of someone's child during the temporary absence of that child's parents'. Accordingly, a babysitter can be defined as a performer of the action of babysitting. The nominal lexeme BABYSITTER is thus semantically more complex than the verbal lexeme BABYSIT and can therefore be regarded as its derivative from a synchronic perspective. (In other words, we can claim that the pair *to babysit–babysitter* is synchronically not different from the pair *to blog–blogger*, in which the latter came into existence via affixation of the former by means of the suffix *-er*.) Similarly, the verbal lexeme PROOFREAD is, from a synchronic point of view, a quasi-idiom whose signified does not only contain the signifieds of the components *proof* and *read* but also the idiomatic signified 'to find mistakes and make corrections': *to proofread* does not mean 'to read proofs' but 'to read proofs in order to find mistakes in them and make the necessary corrections'. Accordingly, a proofreader can be defined as a performer of the action of proofreading. The lexeme PROOFREADER is thus semantically more complex than the lexeme TO PROOFREAD and can therefore be regarded as its derivative from a synchronic point of view.

This does not mean to deny that Marchand's analyses are correct from a diachronic perspective, i.e. as in the case of *to tase* and *skeeve*, immediately after their creation the output verbs *babysit* and *proofread* came to be associated with the quasi-idiomatic signifieds 'to do the job of a babysitter' and 'to do the job of a proofreader'. However, this is no longer true of present-day English speakers who do not remember the true etymologies of these verbs.

Word-formation: basic issues

To conclude, the diachronic history of the lexeme under analysis (its true etymology) may be at odds with present-day speakers' intuitions (its folk etymologies). Accordingly, a student of lexeme-formation must always specify whether he or she approaches lexeme-formation from a synchronic or a diachronic perspective. (The present textbook does not argue for either the former of the latter approach: both can provide important insights into the history and nature of morphological processes that produce new lexemes.)

4.2.5 Why do speakers of English create new lexemes?

The most important reasons are as follows.

1. **lexical gaps**
2. **taboo**
3. **language users' desire for expressivity**

With regard to lexical gaps, consider the lexeme FAKE-ATION 'a vacation where a significant amount of time is spent reading email and performing other work-related tasks' (Word Spy). According to Word Spy, FAKE-ATION is a relatively new lexeme in English: its earliest citation provided by the database dates February 16, 2009. From a formal point of view, the lex *fake-ation* can be analyzed as a blend of the components *fake* and *vacation*: *fake-ation* = *fake* *vacation*. From a semantic point of view, the lexeme FAKE-ATION can be analyzed as a quasi-idiom whose signified does not only contain the signifieds 'fake' and 'vacation', inherent in the blended components *fake* and *vacation*, but also the idiomatic signified 'defining characteristics of such a fake vacation distinguishing it from other fake vacations', i.e. a fake-ation is a particular kind of fake vacation.

It is evident that the blend *fake-ation* appeared in English in the late 2000s (but not in, say, the 1950s) because the phenomenon of a fake-ation is a new phenomenon, brought about by the development of the Internet during the last two decades. Apart from FAKE-ATION, there are many other lexemes whose creation in the 1990s-2000s can be attributed to the need to fill the lexical gap, resulting from the development of the Internet. For example, the quasi-idiom BROWSER 'a computer program enabling Internet users to browse Internet pages', the semi-idiom SEARCH ENGINE 'a service like Google or Bing that allows Internet users to search for information on the Internet', the full-idiom FIREWALL 'a software protecting your computer from hackers' attacks and viruses', etc.

In addition to lexical gaps, new lexemes often come into existence because of taboo. In linguistics, the term 'taboo' is used in connection with unpleasant topics such as, for example, death, illnesses and disabilities, poverty, urination and defecation, sexuality, etc. These topics are said to be **marked by taboo**, which means that we do not talk about these topics as freely as we talk about **taboo-free topics** such as, for example, watching TV, eating, sleeping, washing hands, etc. However, on some occasions we cannot avoid talking about things like having sexual relationships with other people, going to the toilet, being seriously ill, dying, and so on. For instance, many people have sexual partners whom they are not officially married to. To describe the relationship holding between these people without explicitly referring to sex, speakers of English usually use the fully-idiomatic lexemes BOYFRIEND$_2$ and GIRLFRIEND$_2$.

Lexemes like BOYFRIEND$_2$ and GIRLFRIEND$_2$ are usually called **euphemisms**. A euphemism is a lexeme whose lex 'indirectly' expresses a taboo-marked concept. For example, by calling somebody *boyfriend*, we do not directly name the most important but taboo-marked characteristic of the relationship holding between that person and his girlfriend, i.e. that the two are sexual partners. Instead, we use a peripheral but taboo-free characteristic 'friendship', which is sometimes (but not always) typical of the boyfriend–girlfriend relationship.

The need to create a euphemism can thus be considered another important motivation for lexeme-creation. However, it must be observed that euphemistic expressions are not necessarily lexes which realize new lexemes. Thus the famous American euphemisms *bathroom* and *restroom* (when used to refer to toilets rather than to rooms where people can take a bath / have a rest) can be said to be associated with the signified 'toilet' and can therefore be regarded as allolexes of the already existing TOILET lexeme rather than as lexes realizing new lexemes. Hence the need to create a euphemism is usually a motivation for lex-formation rather than for lexeme-formation.

Another reason accounting for the creation of new lexemes, which is far more important than both lexical gaps and taboo, is language users' expressivity or creativity. As was pointed out by Zipf (1949: 19), "the main motivation for speaking is to achieve success". Achieving success by means of speaking does not really mean to be understood by other people. Much more important than this is what Keller and Kirschbaum (2003: 12) call the wish to **produce an impression on other people** ("der Wunsch zu imponieren oder positiv aufzufallen"), the wish to show off. To illustrate this point, let us consider the lexeme FLUNAMI 'an overwhelming number of flu cases in the same area at the same time' (Word Spy). From a formal point of view, the lex *flunami* can be analyzed as a blend of *flu* and *tsunami*. From a semantic point of view, the lexeme FLUNAMI can be analyzed as a semi-idiom whose signified contains the signified of the component *flu* but not of the shortened component *-nami* (which is a morfoid whose signified 'tsunami' partially motivates the idiomatic meaning 'flunami'). Just like FAKE-ATION, FLUNAMI is a relatively new lexeme in English: its earliest citation provided by Word Spy dates January 04, 2006. However, in contrast to the phenomenon of a fake vacation that involves reading non-private emails while on vacation, the phenomenon of a flunami is definitely not a new phenomenon: influenza epidemics had surely occurred long before the mid-2000s. Accordingly, the creation of FLUNAMI cannot be attributed to the urgent need to fill the lexical gap, arising from the emergence of a new phenomenon. Also, it cannot be attributed to the need to create a euphemism: influenza is not a taboo-marked illness. The creation of FLUNAMI can only be due to the above named desire for creativity as a means of achieving success. Thus the creator of *flunami* demonstrates his or her creativity in that he or she invents a semi-idiomatic blend one of whose components (*-nami*) is a morfoid whose signified partially motivates the idiomatic meaning of the blend. The same can be said about the creator of *fake-ation*. In addition to inventing a new lex for expressing a supposedly new phenomenon, he or she demonstrates his or her ability to creatively use the available lexeme-building resources of the English language, thereby producing an impression on other people.

Finally, it must be mentioned that the creators of lexemes like FAKE-ATION and FLUNAMI achieve success because of the **hypostatization potential** of words. A well-known fact is that "the existence of a particular word creates the impression that there is a corresponding thing or entity to which the word refers to" (Schmid 2008: 5). This effect is usually called **hypostatization**. Indeed, the existence of the blend *fake-ation* creates the impression that there are masses of people who spend their vacations answering non-private emails, even though we do not know for sure whether this is actually the case. Undoubtedly, there are many people who check their emails while on vacation, but it is not clear whether because of this, they regard their vacations as fake vacations. The term *fake-ation* does, however, create the impression that fake-ations constitute a serious problem which negatively affects the lives of many people.

Precisely because of the hypostatization potential of words, new lexemes may be created in an attempt to draw other people's attention to phenomena like fake-ation which the creators of these lexemes find particularly important. Consider, for example, the following lexemes.

- PRECARIAT 'people whose lives are precarious because they have little or no job security' (Word Spy)

- ON-CALL-OGIST 'a doctor who is frequently on call, particularly one who earns a living by filling in for other doctors' (Word Spy)

- BIKELASH 'a strong, negative reaction towards cyclists, particularly by police officers or drivers' (Word Spy)

Again, as in the case of the lexeme FAKE-ATION, the mere existence of these lexemes creates the impression that there are many people whose lives are precarious because of the absence of job security; there are many doctors who earn a living by filling in for other doctors; there are many police officers and drivers who are aggressive towards cyclists. The creators of these lexemes thus wanted to achieve success by drawing other people's attention to (in their view) important social problems.

4.2.6 The establishment of new lexemes

As our starting point, let us compare the synchronic status of the lexemes BOYFRIEND₂ and FAKE-ATION. The obvious difference is that the former is an **established lexeme** (i.e. one which is perceived to be the norm of the English language), whereas the latter is a **neologism**, i.e. a lexeme that has only recently been created and, accordingly, has not yet become a standard English lexeme.

According to Schmid (2008: 3), the **establishment** of a lexeme – i.e. its development from one speaker's **coinage** to a standard word recognized and used by the majority of members of the same linguistic community – has three stages:

1. **creation**
2. **consolidation**
3. **establishing**

Each of these stages has three perspectives:

1. **structural perspectives**
2. **socio-pragmatic perspectives**
3. **cognitive perspectives**

Let us discuss each of these stages and perspectives using the neologism FAKE-ATION as an illustrative example. We will begin with the structural perspective.

At the moment when some speaker of English produced the blend *fake-ation*, a **nonce-formation** was created. What defines the **creation stage** from a structural perspective is an instability of both the signifier and the signified of a nonce-formation. For example, as regards *fake-ation*, it is not entirely clear how the word must be spelled. Word Spy mentions four distinct orthographic variants: *fake-ation, fakeation, fakation,* and *facation*. The same is true of its signified: according to Word Spy, "*fake-ation* has as many meanings as it does spellings". In addition to the signified discussed above, these include:

- 'calling in sick when you're healthy'
- 'a miserable or problem-filled vacation'

- 'a pretend vacation where you stay at home but take steps to make it appear as [though] you went away (e.g. applying tanning cream)'

Observe that, according to Word Spy, the signified 'a pretend vacation' is the earliest recorded sense of *fake-ation*, which has been documented since 2004.

The next stage – the **consolidation** – is characterized by the beginning of the stabilization of both the signifier and the signified of a former nonce-formation. Thus it appears that despite the still existing variation between the orthographic forms *fake-ation, fakeation, fakation, facation* and the four different signifieds mentioned above, both the signifier and the signified of the lexeme FAKE-ATION have begun to stabilize. We are justified in arriving at this conclusion because Word Spy is not the only Internet source where the lex of this neologism is spelled *fake-ation* and defined as a vacation for people who read non-private emails. Consider, for example, (72).

(72)   *Florida hotel offers 'fake-ation' package to working travelers. People who find themselves taking 'fake-ations' – a new term coined to describe trips taken by those who work during their vacations - may want to take advantage of a new package offered by a Florida resort, reports the Los Angeles Times.* (http://tinyurl.com/67agrll)

The final stage – the **establishment** – which FAKE-ATION apparently has not yet reached, will result in the **full-lexicalization** of a former nonce-formation. That is, there will be a conventionalized association between a particular form and a particular meaning: the signifier *fake-ation* will mean approximately the same thing for the majority of English speakers. (It is also possible that at the end of the consolidation process there will be the vocable FAKE-ATION consisting of several lexemes expressing polysemous signifieds rather than a single FAKE-ATION lexeme.)

Now, let us proceed to the socio-pragmatic perspective. As said above, at the moment when some speaker of English produced the blend *fake-ation*, a nonce-formation was formed, which with the course of time began to **spread** in the English linguistic community. As a consequence of this, *fake-ation* is at present no longer a nonce-formation, i.e. there are many speakers who use this blend or at least understand what it means. (If this were not the case, it would not appear in neologisms databases like Word Spy). However, despite its spreading, *fake-ation* has not yet achieved the status of an **institutionalized lexeme**, i.e. one which is perceived as a standard English lexeme. This can be corroborated by the fact that recent uses of *fake-ation* in contexts like that of (72) are still accompanied by **meta-linguistic comments** concerning both its signified and its socio-pragmatic status, i.e. *'fake-ations' – a new term coined to describe trips taken by those who work during their vacations.* Lexes which realize institutionalized and fully-lexicalized lexemes like BOYFRIEND₂ are usually not accompanied by comments of this type.

Finally, from the **cognitive perspective**, the creation of the new lexeme FAKE-ATION began with the creation of a new concept of a fake-ation. This does not mean that speakers of English had been unaware of the phenomenon of a vacation that involves spending a significant amount of time answering non-private emails before the blend *fake-action* appeared in the late 2000s. However, this phenomenon was not experienced by them as a "manifestation [...] of recurrent and familiar events or personal habits" (Schmid 2008: 8). In other words, despite the existence of the phenomenon of a fake-ation, speakers of English did not have the concept of a fake-ation prior to the creation of the blend *fake-ation* around the year 2009. Accordingly, the coinage of the blend *fake-ation* by a particular speaker of English gave rise to a **pseudo-concept**, i.e. a concept that existed only in the mind of that particular speaker. Then, as the blend began to spread in the English community, the concept of a fake-ation began to hypostatize. That is, more and more people began to experience the phenomenon of such a vacation as a manifestation of a recurrent and a familiar event. This is supported by (72), which mentions that Florida hotel has recently begun offering their guests the so-called Perfect 'Fake-ation' package. According to the hotel Web site,

> The Shores Resort & Spa has created a new package for career-conscious travelers who want to stay connected to work while on vacation – the 'Fake-ation' Package. According to TripAdvisor®, 59 percent of travelers are connected to work than more ever before during leisure travel, with 62 percent checking work email, and 13 percent calling into the office while on vacation. Starting at just $119 per night, the new Fake-ation package caters to this growing group of business / pleasure travelers, with WiFi throughout the resort – including the pool area – a business center available 24-hours daily, coffee on demand, the Wall Street Journal upon check-in, and a private office available for conference calls. Call 866.934.7467 to book your 'Fake-ation' today! (http://tinyurl.com/ddbgqs).

Obviously, the existence of such a package demonstrates that the concept FAKE-ATION is already a fairly hypostatized concept.

4.2.7 The non-institutionalization of new lexemes

Not all nonce-formations develop into fully-established lexemes. The reasons for this are usually of an **extra-linguistic character**. That is, for example, if we suddenly stop checking emails while on vacation, the neologism FAKE-ATION will most likely not become an institutionalized lexeme. Also, it seems that PRECARIAT has more chances to fully-institutionalize than BIKELASH: many people do not have job securities and are therefore indeed in a rather precarious situation. By contrast, the lexeme BIKELASH seems to denote a less significant social problem: there are many people who are neither cyclists nor police officers and hence do not really know (or do not care much about the fact) that the latter are aggressive towards the former.

In addition to extra-linguistic reasons like those named above, the non-establishment can be due to (linguistic) **blocking**. As an illustration, let us consider the non-institutionalized formations *studier* and *liver*. According to the OED, the signifier *studier* denoting 'a person who is addicted to or engaged in study; a student' appeared in English around 1380. By 1593 the lexeme STUDIER$_1$ 'a person engaged in study / student' had given rise to the quasi-idiom STUDIER$_2$ 'one who studies a specified subject'. Finally, around 1597 there emerged the full-idiom STUDIER$_3$ 'one who strives after or pursues (an object or end)'. Of these lexemes, none can be regarded as an established lexeme of Present-day English. According to the OED, STUDIER$_1$ is **obsolete** (i.e. non-existent in Present-day English), whereas STUDIER$_2$ and STUDIER$_3$ are marked as rare.

A slightly different situation is characteristic of *liver*. According to the OED, the non-idiomatic lexeme LIVER$_1$ 'a person who lives or is alive, a living creature' was created in 1382. This lexeme still exists in Present-day English. However, according to the OED, it can be found only in the South and the South-West of England. Apart from LIVER$_1$, there are also the quasi-idiomatic lexemes LIVER$_2$ 'an inhabitant, a dweller' (i.e. a person who lives at a particular place) and LIVER$_3$ 'a person who lives a life of pleasure or activity'. Of these lexemes, only LIVER$_3$ can be regarded as a more or less established lexeme of Present-day English, whereas LIVER$_2$ 'an inhabitant, a dweller' occurs only in the U.S.A.

In summary, what both the vocables **STUDIER** and **LIVER** seem to share is that their non-idiomatic lexemes STUDIER$_1$ 'a person who studies' and LIVER$_1$ 'a person who lives', which existed in English in the past, did not become fully-institutionalized lexemes of Present-day English. That is, we cannot use the affixed forms *studier* and *liver* to refer to performers of the actions of studying and living. In this respect, *studier* and *liver* are different from forms like *baker, teacher, preacher, worker, writer*, etc.

The non-institutionalization of the lexemes STUDIER$_1$ and LIVER$_1$ is often attributed to blocking. As for *studier*, it is argued that its establishment was prevented by the synonymic lexeme STUDENT. A very similar case is an often cited non-institutionalized formation *stealer*. According to the OED, the lexeme STEALER$_1$ 'one who steals; a thief' appeared in English around 1508. However, just like STUDIER$_1$, it did not manage to develop into a fully-established lexeme. As we pointed out in 2.4.7, in Present-day English there is only the quasi-idiomatic lexeme STEALER₂ 'one who steals something specified'. A possible explanation for the non-institutionalization of STEALER₁ is the existence of the fully-synonymic lexeme THIEF. (If we accept these analyses, we can regard the non-institutionalization of STUDIER₁ and STEALER₁ as instances of **synonymic blocking**.)

In contrast, the non-institutionalization of the lexeme LIVER₁ is usually attributed to **homonymic** and **semantic blocking**. As for the former, the signifier *liver* is associated with the totally unrelated homonymic meaning 'an inner organ of the human body'. According to the OED, the lexeme LIVER 'an inner organ' has existed since the Old English period, whereas LIVER 'a person who lives' did not appear until the end of the 14th century. It is possible that the former could have blocked the institutionalization of the latter. As argued by Ullmann (1957: 122), a **homonymic clash** like that of LIVER 'an inner organ' and LIVER 'a person who lives' is a "pathological situation [which] arises whenever two or more incompatible senses [...] develop around the same name".

At the same time, however, as Plag (2003: 64) argues, this homonymic clash could not have been the sole reason for the non-institutionalization of LIVER₁. Indeed, all languages tolerate homonyms. For example, as was observed in 2.2.1, in English there are the homonymic lexemes CASE 'a set of circumstances or conditions' and CASE 'a box or receptacle for holding something'. Similarly, there are the homonymic lexemes BANK 'financial institution' and BANK 'the ground near the river'. If homonymic clashes were indeed such an intolerable pathology, the just named pairs of homonymous lexemes (both of whose members are fully-institutionalized lexemes in Present-day English) would most likely not exist.

According to Plag, the non-institutionalization of LIVER₁ can be attributed to the fact that the derivational suffix *-er* typically denotes an **agent**, i.e. a deliberate initiator of the action specified by a preceding verb. That is, a preacher is the agent of the action of preaching (a person who deliberately preaches); a teacher is the agent of the action of teaching (a person who deliberately teaches); a writer is the agent of writing (a person who deliberately writes); etc. The agentive meaning is also evident in formations like *New Yorker* 'an inhabitant of New York', *Londoner* 'an inhabitant of London', *Berliner* 'an inhabitant of Berlin', etc. A New Yorker is the agent of the action of living in New York: he or she initiated this action by either moving to New York or (if he or she was born in New York) by not moving to a different place. By contrast, living (in the sense 'just being alive') is not an agentive activity. We are not the agents but the **experiencers** of the action of living. We live as long as we are alive. This is an inherent property of all living creatures rather than a deliberately initiated activity.

That this could indeed have been the reason for the non-institutionalization of LIVER₁ is corroborated by the existence of the lexeme LIVER₃ 'a person who lives a life of pleasure or activity'. (According to the OED, this lexeme is neither obsolete nor rare and can thus be regarded as more institutionalized than LIVER₁.) Thus a person who lives a life of pleasure or activity is obviously the agent of the action of living a life of pleasure or activity: he lives such a life because he decided to do so.

A number of recent studies regard blocking not (only) as a diachronic phenomenon accounting for the non-institutionalization of a particular lexeme (which, however, existed at some point in the past) but rather as a synchronic explanation for the non-occurrence of new formations like STEALER₁ and STUDIER₁. In other words, why do present-day English speakers (who do not remember that STEALER₁ and STUDIER₁ existed in the past but could not develop into fully-institutionalized lexemes and therefore disappeared from the English language) not create these lexemes anew? Or to be more precise, why do present-day speakers of English not create additional allolexes of the lexemes THIEF and STUDENT?

The synchronic answer to this question given by these studies is, however, identical with what was said in connection with the diachronic perspective discussed above: the formation of the allolexes *stealer* and *studier* is still blocked by the established and frequently used lexes *thief* and *student*. As e.g. Rainer (2005: 337) argues, the degree of synchronic semantic blocking largely depends on the frequency of the blocking lex: frequently used lexes have a stronger blocking potential than low frequency lexes; e.g. *thief* and *student* are still capable of blocking the formation of the allolexes *stealer* and *studier* because present-day English speakers fairly often hear the words *thief* and *student*.

This textbook cannot accept this analysis. As we have seen at earlier points, English (as well as any other language) abounds in lexemes whose lexes express fully- or at least near-synonymic signifieds. For example, *true* and *faithful* are fully-synonymic allolexes of the same TRUE / FAITHFUL lexeme; FEVER and PYREXIA are fully-synonymic allolexes of the same FEVER / PYREXIA lexeme; BATHROOM and RESTROOM are fully-synonymic allolexes of the same TOILET lexeme; etc. Moreover, as we will see in the next section of this chapter, English has several mechanisms that are exclusively used for lex-formation (i.e. the formation of fully-synonymic allolexes of already existing lexes). One of such mechanisms is **clipping**, i.e. the shortening of an input lex. For example, *girlf* is a fully-synonymic clipped allolex of the input lex *girlfriend*. According to the OED, *girlf* has been recorded in (chiefly British) English since 1991. Note that the input lex *thief*, which supposedly blocks the formation of the synonymous allolex *stealer*, has a lower frequency of use than *girlfriend*, which has recently served as an input lex for the synonymous allolex *girlf*. BYU-BNC has 1332 occurrences of *girlfriend*, but only 730 occurrences of *thief*. Why is it then that the more frequently used lex *girlfriend* did not block the formation of the fully-synonymous allolex *girlf*, whereas the less frequently used lex *thief* still blocks the formation of the fully-synonymous allolex *stealer*?

The answer to this question seems to be that the clipping *girlfriend* → *girlf* is an instance of deliberate lex-formation, i.e. the clipper of *girlfriend* wanted to have a shorter allolex for expressing the same signified 'girlfriend'. The creator of *girlf* knew from the very beginning that the clipped output lex would become a fully-synonymic allolex of the corresponding non-clipped input lex. Also, he or she knew that the two allolexes would occur in stylistically relevant free variation: i.e. that the shorter *girlf* would be a less formal variant of the longer *girlfriend*. (This is a general characteristic of clipping: clipped output lexes are usually less formal than corresponding non-clipped input lexes.) By contrast, if someone ever coins *stealer* and *studier*, this will most likely not be because of the wish to create additional allolexes of the established lexes *thief* and *student* but because of the wish to form affixed forms by means of the derivational suffix *-er*. The consequence of this, however, will be the formation of two additional allolexes of uncertain status. This means that it will not be clear whether *thief* and *student* must retain their status of the primary signifiers of the lexemes THIEF and STUDENT or instead be superseded by the new lexes *stealer* and *studier*. The latter lexes will thus 'encroach' on the dominant status of the former, thereby considerably enhancing the possibility of blocking (or to be more precise, the possibility of the non-institutionalization) of *stealer* and *studier*.

At the same time, note that there is absolutely no reason why *thief* and *student* must block the formation of stylistically different allolexes *stealer* and *studier*. That is, if some speaker of English decides to use these affixed forms as less formal allolexes of *thief* and *student*, there will be no synonymic blocking: if the lexes *stealer* and *studier* coexisted with *thief* and *student* in the past, it is very possible that this situation will recur in the future.

Finally, let us briefly return to the complementary distribution of the allomorphs *un-* and *in-*, which we discussed in Chapter 2. As was observed in 2.3, *in-* combines only with words of Latin or Romanic origin, whereas *un-* is used with native or completely naturalized words. This is the reason why there can only be *untrue* and *inadequate* but not **intrue* and **unadequate*. For some authors, examples like this are instances of **morphological blocking**.

This book argues for a different terminological solution. First of all, notice that like synonymic blocking, so-called morphological blocking exemplified by the impossibility of **intrue* and **inadequate* seems to be more relevant for lex-formation rather than for lexeme-formation. That is, if it were not for these combinatory restrictions, it would now be possible to form additional allolexes of the lexemes UNTRUE and INADEQUATE. Second, as was stated in 2.3, the impossibility of the formations **intrue* and **unadequate* is due to the different syntactics of the prefixes *un-* and *in-*: there can be no **intrue* because the syntactics of *in-* does not allow us to combine this prefix with the Germanic input lex *true*. And there can be no **unadequate* because the syntactics of *un-* does not allow us to combine this prefix with the Latinate input lex *adequate*. This phenomenon could perhaps be better referred to as **syntactics' blocking** rather than morphological blocking. However, observe that in contrast to the examples discussed above, in the case of syntactics' blocking, there is no prevention of the institutionalization of either a new lexeme or a new lex of the same lexeme: neither the former nor the latter is actually formed as a result of existing syntactics' blocking. That is, according to the OED, the lex **intrue* has never existed in English.

Taking this into account, this book suggests that the term 'blocking' should be reserved only for cases of semantic blocking in which the institutionalization of an already formed lexeme or a lex of the same lexeme is prevented by some inherent semantic properties of the signified of the lexeme in question, whereas the non-existence of forms like **intrue* and **unadequate* should be regarded as consequences of the inherent properties of the syntactics of the prefixes *un-* and *in-* rather than as instances of blocking.

4.2.8 Productivity

The term **'productivity'** refers to the ability of a particular lex- or lexeme-building mechanism to produce new signifiers, be it lexes that realize new lexemes or additional allolexes that realize already existing lexemes. In the case of affixation, we usually speak of the productivity of an affix (e.g. the productivity of the derivational suffix *-er*), whereas in the case of processes like compounding, conversion, etc., we speak of the productivity of a particular formation pattern (e.g. the productivity of the noun → verb conversion).

Since productive morphological processes are those that give rise to new signifiers, any study of productivity must begin with the identification and documentation of recently coined signifiers. With regard to lexeme-formation processes, we can rely on databases of English neologisms such as, for example, the already mentioned Word Spy (http://www.Word Spy.com/).

Word Spy is a continually updated, online-based database of neologisms that appeared in the English language during the last three decades. The entries of the database are arranged both chronologically and thematically. However, if you click at 'Posts by date' and then at e.g. '2011', Word Spy will list all neologisms that were added to the database in the year 2011, not the neologisms that were necessarily created in 2011. For example, the lexeme COPYFIGHTER 'a person who opposes copyright laws and practices that he or she perceives to be unfair' was posted to the Word Spy Web site in 2011. However, its earliest citation provided by Word Spy dates January 22, 2003. Similarly, there is an entry for the lexeme AFTER-PARTY 'a social gathering that occurs after a party, especially after a rave' which was added to the database in 2002. However, its earliest citation dates March 17, 1980.

If you click at 'Posts by category', Word Spy will list a number of (mainly thematic) categories into which recently created lexemes have been classified by the author of the Web site. These include, for example, 'Books and magazines', 'Corporate culture', 'Economics', 'Health and fitness', 'Money', 'Medicine', 'Psychology', 'Race', etc. If you click at e.g. 'Money', Word Spy will list all recently created lexemes which, according to the author of the Web site, have something to do with money.

Note that some of the categories are not thematic but genuinely linguistic categories. For example, there is the category 'Verbed nouns', listing verbal lexemes which have recently come into existence via morphological conversion of corresponding input nominal lexemes. For example:

- TO BACKGROUND 'to surreptitiously perform a task in the background while one's attention is supposed to be on another task' (Word Spy)

- TO OFFICE 'to perform office-related tasks, such as photocopying and faxing' (Word Spy)

- TO PIE 'to hit a person, particularly a political or business leader, in the face with a pie' (Word Spy)

Also, there is the category 'Idioms', which lists recently coined idiomatic phrases. For example:

- BIRDS OF A FEATHER MEETING 'a meeting held at a computer-related trade show or conference in which people who work in the same technology area at different companies exchange information and experiences' (Word Spy)

- NAILING JELLY TO A TREE 'tackling a particularly messy, and probably impossible, problem' (Word Spy)

- TO PUT WOOD BEHIND THE ARROW 'to provide a product or company with money and other resources' (Word Spy)

If you want to study the productivity of a derivational affix, go to the Advanced search page at http://wordspy.com/search.asp, enter '*suffix' or 'prefix*' (both without quotation marks) to the search mask, choose the option 'Within title', and then click at 'Search'. That is, for example, if you want to learn whether Word Spy contains neologisms whose lexes came into existence via affixation by means of the verb-forming suffix -*ize* (e.g. *centralize*), enter '*ize' (without quotation marks). If you want to learn whether Word Spy contains neologisms whose lexes came into existence via affixation by means of the prefix *un-*, enter 'un*' (again without quotation marks). Word Spy will then yield all words ending in -*ize* and beginning with *un-*. Some of these words could be verbs containing the suffix -*ize* and negative adjectives containing the prefix *un-*. As regards the former, Word Spy contains, for example, the following verbs:

- TO BAGONIZE 'to wait anxiously for your suitcase to appear on the airport baggage carousel' (Word Spy)

- TO DOLLARIZE 'for a country to abandon its national currency in favor of the U.S. dollar' (Word Spy)

- TO VELOCITIZE 'to cause a person to become used to a fast speed' (Word Spy)

As regards the latter, Word Spy contains the already mentioned negative adjective *ungoogleable*. Accordingly, we are justified in concluding that both the suffix -*ize* and the prefix *un-* are productive derivational affixes in Present-day English.

An alternative to Word Spy is the OED, or to be more precise, the online version of the OED available at http://www.oed.com/. Note that you or your institution must subscribe to the electronic version of the OED in order to be able to access it online.

The two major advantages of the OED is that it enables us to (easily) determine the productivity of virtually any lexeme-building mechanism as well as to compare the **diachronic productivity** of a particular mechanism. For example, if you want to find out (using Word Spy) whether speakers of English have recently created new lexemes by means of apophony, you have to manually analyze each of the Word Spy entries: at present, it is impossible to order Word Spy to specifically search for instances of apophony (as well as for instances of such lexeme-building mechanisms as lexeme-manufacturing, back-formation, borrowing, etc.). By contrast, using http://www.oed.com/, you can search the OED for instances of all these mechanisms. Just go to the OED Advanced search page at http://www.oed.com/advancedsearch, enter 'alteration' (without quotation marks) to the search mask of the first row from above, choose the option 'Etymology' instead of 'Full text' (near to the search mask), enter '1990-' to the 'Date of entry' below the search mask, and then click at 'Search'. The OED will then find all recent words (i.e. those which appeared after 1990) whose Etymology-section contains the word *alteration*, i.e. how the OED calls phonetic alterations similar to those of *to increase* → *an increase* and *bouldering* → *buildering*, which we discussed in 4.2.2. Most of the results will be forms that came into existence via apophony.

Recent instances of apophony provided by the OED include, for example:

- *adultescent* 'an adult who has retained the interests, behavior, or lifestyle of adolescence' (OED / 1996) ← *adolescent*

- *Google* 'a proprietary name for an Internet search engine launched in September 1998' (OED / 1999) ← *googol* 'a fanciful name [...] for ten raised to the hundredth power' (OED / 1940)

- *Lollywood* 'The Pakistani film industry, based in Lahore [...]' (OED / 1995) ← *Bollywood* 'the Indian film industry, based in Bombay' (OED / 1976)

Given these examples, we can claim that apophony is still a productive lexeme-building mechanism in Present-day English.

If you are interested in the diachronic productivity of apophony (i.e. the question whether this mechanism was more / less productive during a particular period in the history of the English language), repeat the same search without entering '1990-' to 'Date of entry'. The OED will then find all instances of apophony that have ever been created in English. Then click at 'Timeline' (near 'View as: List') and the OED will yield a graph consisting of 20 bars. Each bar corresponds to the interval of 50 years. The higher the bar is, the larger the number of words that came into existence via apophony during that particular interval of time.

Finally, it is often suggested that the synchronic productivity of a lexeme-building mechanism (especially, the productivity of a particular derivational affix) can be determined with the help of a synchronic corpus like COCA. The point here is that neologisms can often be found among low frequency forms, in particular, among so-called **hapax legomena**, i.e. forms which occur no more than once in a given corpus. The reason for this is obvious: (lexes which realize) newly created lexemes which have not yet reached the status of fully-established lexemes are usually not used very often.

For example, if you want to determine the synchronic productivity of the noun-building suffix *-hood* (of e.g. *childhood* and *waithood*) using COCA, you have to enter '*hood.[nn*]' (without quotation marks) to the search mask of the corpus and click at 'Search'. COCA will then yield all nouns (occurring in this corpus) that end in *-hood*. Some of them will be suffixed nouns like *childhood* and *waithood*.

Then you have to write down all hapax legomena yielded by COCA and check whether these forms are indeed lexes that realize previously unattested lexemes. (This is necessary because a hapax legomenon is not necessarily a neologism. It may be just a low frequency word that appeared long ago.) For this purpose, establish whether the hapax legomena found by COCA occur in dictionaries like the OED or MWO. If you succeed in finding at least one **unattested hapax** (i.e. a form for which neither the OED nor MWO have an entry), you will be justified in claiming that the derivational affix under analysis is still a productive noun-building affix.

At present, COCA does indeed contain a number of hapax legomena that are not listed in either the OED or MWO. These include, for example, *refugee-hood*, *rookiehood*, *otherhood*, *pointhood*, etc. Consider, for instance, the meaning of *refugee-hood* in (73).

(73) *The fact that Palestinian cultural and political communities have not yet coincided in time and space also points to the current link between Palestinian identity and refugee-hood* (COCA)

It appears that the meaning of the *refugee-hood* of (73) is very similar to that of *childhood* and *waithood*. Like the latter, *refugee-hood* also seems to denote the state of being a refugee. Accordingly, we can conclude that *-hood* is still a productive suffix that attaches to nouns like *refugee*, thereby forming nouns denoting a state: *refugee-hood* = the state of being a refugee.

If you want to establish the synchronic productivity of a prefix like *un-*, just enter 'un*.[j*]' (without quotation marks) and repeat the procedure discussed above. (COCA will then search for adjectives beginning with *un-*. If you enter only 'un*', COCA will search for lexes of all word classes beginning with *un-*.)

4.3 Lex-formation

As we argued in 4.1, lex-formation is different from lexeme-formation in that it produces allolexes of already existing lexemes rather than lexes realizing new lexemes. Another important difference is that lex-formation is never caused by the need to fill a lexical gap: this is a priori impossible, given that lex-formation never gives rise to new lexemes. The primary motivation for lex-formation is the language user's desire to show off, the wish to produce an impression on other language users by replacing an old 'dull' lex by a new 'fresh' allolex.

Below we will discuss the most important mechanisms that speakers of English have at their disposal for creating allolexes of already existing lexemes. These mechanisms include:

1. **lex-forming clipping**
2. **lex-forming suppletion**
3. **lex-forming abbreviation**

4. lex-forming borrowing
5. lex-forming apophony
6. lex-forming affixation
7. lex-forming syntactics' change
8. lex-forming orthographic modification

4.3.1 Lex-forming clipping

Lex-forming clipping is the shortening of an input lex. According to Marchand (1969: 442-445), clipped allolexes can be classified into:

1. **back-clippings**
2. **fore-clippings**
3. **mid-clippings**

Back-clippings are clipped output allolexes in which the back part of their non-clipped input lexes is retained. For example, *girlf* (← *girlfriend*), *mobe* (← *mobile*), *refi* (← *refinancing*), etc.

Fore-clippings are clipped output allolexes in which the fore part of their non-clipped input lexes is retained. For example, *brane* (← *membrane*), *droid* (← *android*), *fro* (← *Afro*), etc.

Finally, mid-clippings are clipped output allolexes in which the middle part of their non-clipped input lexes is retained. For example, *flu* (← *influenza*), *fridge* (←*refrigerator*), *jams* (←*pyjamas*), etc. (The last example *jams* can be regarded as both a fore-clipping and a mid-clipping.)

Lex-forming clipping is often a means of creating less formal first names. For instance, *Alex* is a less formal back-clipping of *Alexander*; *Tina* is a less-formal fore-clipping of *Christina*; *Liz* is a less formal mid-clipping of *Elizabeth*; etc.

4.3.2 Lex-forming suppletion

In addition to clipping, informal first names can be formed with the help of suppletion. For example, in Russian *Sasha* is a less formal suppletive allolex of the first names *Alexander* and *Alexandra* (which are of Greek origin). In the English linguistic community, the signifiers *Sasha* and *Alexander* are usually regarded as two different names (see e.g. Hanks et al. 2006) and hence must be seen as lexes realizing two different lexemes; e.g. the British comedian *Sacha*[9]

---

[9] *Sacha* is a spelling variant of *Sasha*.

Baron Cohen cannot be addressed as *Alexander Baron Cohen*. At the same time, observe that the American skater Alexandra Pauline Cohen is usually referred to as *Sasha Cohen*. (The reason for this is, however, the Russian origin of her mother, who emigrated to the U.S.A. from the Soviet Union.[10]). Accordingly, at least in this case, we can analyze the signifier *Sasha* as a less formal suppletive allolex of *Alexandra*.

Genuinely English instances of lex-forming suppletion include, for example, *Bill* ← *William*, *Bob* ← *Robert*, *Dick* ← *Richard*, *Ted* ← *Edward*, etc.

4.3.3 Lex-forming abbreviation

Lex-forming abbreviation can be illustrated by the shortening of input idiomatic phrases like *British Broadcasting Corporation* and *Oxford English Dictionary* to shorter output allolexes like *BBC* and *OED*. That is, both the former and the latter are quasi-idioms, whose signifieds are not entirely representable in terms of their components' signifieds: both *British Broadcasting Corporation* and *BBC* denote a particular corporation that broadcasts in Britain and both *Oxford English Dictionary* and *OED* denote a particular English dictionary that has something to do with Oxford.

Lex-forming abbreviation must be distinguished from (lexeme-building) blending. Compare, for example, the just named abbreviations *BBC* / *OED* and the blend *BRICs* 'the countries of Brazil, Russia, India, and China viewed as a group of emerging economies with large potential markets' (Word Spy). At first glance, it may seem that *BRICs* came into existence in essentially the same way as both *BBC* / *OED*. That is, *BBC* = ~~British Broadcasting Corporation~~ and *OED* = ~~Oxford English Dictionary~~. Similarly, *BRICs* = ~~Brazil, Russia, India, China~~. Nevertheless, despite this formal similarity, *BBC* / *OED* and *BRICs* must be analyzed as instances of two different processes: lex-forming abbreviation and lexeme-building blending. The difference between the former and the latter is that while the shortening of *British Broadcasting Corporation* and *Oxford English Dictionary* to *BBC* and *OED* gave rise to allolexes of already existing lexemes – as we said above, *BBC* and *OED* mean the same thing as *British Broadcasting Corporation* and *Oxford English Dictionary* – the shortening of *Brazil, Russia, India, China* to *BRICs* gave rise to a new quasi-idiomatic lexeme whose signified is not entirely representable in terms of its components' signifieds: *BRICs* does not mean 'Brazil' + 'Russia' + 'India' + 'China' but 'these countries viewed as an economic unit'. Accordingly, *BRICs* is an instance of an anisomorphic lexeme-building blending, whereas *BBC* and *OED* are products of lex-building abbreviation.

---

[10] See http://tinyurl.com/ygfwam8.

Abbreviations are traditionally classified into **alphabetisms** and **acronyms**. The former are abbreviations that are pronounced letter by letter. For instance, *BBC* /biːbiːˈsiː/. The latter are abbreviations that are pronounced as normal words. For example, *SARS* (← ~~severe acute respiratory syndrome~~) is pronounced /sɑːz/. Very often one and the same abbreviation can be pronounced as both an alphabetism and as an acronym. For example, *FAQ* (← ~~frequently-asked questions~~) is pronounced both /ɛfeɪˈkjuː/ and /fak/.

4.3.4 Lex-forming borrowing

Lex-forming borrowing is the process of borrowing a foreign language signifier with the purpose of creating a stylistically different allolex of a lex of an already existing lexeme. For example, the borrowing of the Latin *pyrexia*, which took place around 1777, gave rise to a stylistically different allolex of the already existing lex *fever*, which has been recorded in English since circa 1000 (OED). Similarly, the borrowing of the German signifier *hausfrau* in 1798 gave rise to an additional allolex of the already existing lexeme HOUSEWIFE 'a woman (usually, a married woman) who manages or directs the affairs of her household' (OED), which has been realized by the native English lex *housewife* since circa 1225.

4.3.5 Lex-forming apophony

Lex-forming apophony has recently given rise to the following allolexes:

- *feck* (← *fuck*)
- *Merkin* (← *American*)
- *shedload* 'a large amount or number' (← *shitload*)

As in the case of all other lex-forming mechanisms discussed above, these output signifiers have the same meaning as their corresponding input signifiers: *feck* means 'fuck', *Merkin* means 'American', *shedload* means 'shitload'.

4.3.6 Lex-forming affixation

In Section 4.1 we got acquainted with one instance of lex-forming affixation: the formation of adverbs from adjectives by means of the lex-forming suffix *-ly*. E.g. *happy* → *happily*.

Word-formation: basic issues 107

Apart from this, lex-forming affixation also includes so-called **expletive infixation**. Recall the infix *-ma-* of *edumacation*, which we discussed in 2.5.4. Since the output signifier *edumacation* has a different meaning than the input signifier *education* – *edumacation* means 'poor education' – we are justified in concluding that the infix *-ma-* is a usual lexeme-building affix. Now, consider the infix *-my-* of *jurismydiction* in (74).

(74)   Agent Smith: *Lieutenant.*
       Lieutenant: *Oh shit.*
       Agent Smith: *Lieutenant, you were given specific orders.*
       Lieutenant: *Hey, I'm just doing my job. You give me that juris-<u>my</u>-diction crap, you can cram it up your ass* (The Matrix)

This exchange takes place in the beginning of the famous science fiction film *The Matrix*. Obviously, the infix *-my-* inserted by the Lieutenant to the input signifier *jurisdiction* does not create a lex realizing a different lexeme: the output signifier *juris-my-diction* has exactly the same meaning as the input signifier *jurisdiction* and, accordingly, can be considered its allolex. What *-my-* does here is express the emotional state of the Lieutenant, who is not quite happy with Agent Smith's interference. The infix *-my-*, in contrast to the infix *-ma-* of *edumacation*, is thus a lex-forming affix.

4.3.7 Lex-forming syntactics' change

Lex-forming syntactics' change can be illustrated by means of the syntactic contrast between the *early* of (75) and the *early* of (76).

(75)   *John Andrew's own family – Gala and their two girls – had departed <u>early</u> in May [...]* (COCA)
(76)   *Glancing back at the party on the lawn, he saw a bespectacled woman twirling shirtless on the sidewalk and regretted his <u>early</u> departure* (COCA)

While the *early* of (75) has the syntactics of an adverb (cf. *They lived happily*), the *early* of (76) has the syntactics of an adjective (cf. *his happy life*). This syntactic contrast is, however, not accompanied by a difference in meaning: both the former and the latter carry the meaning 'early'. Accordingly, the *early* of (75) and the *early* of (76) can be regarded as allolexes of the same lexeme which occur in complementary distribution.

Compare also the signifieds and the syntactics of the *sells* of (77) and the *sells* of (78).

(77)   The book <u>sells</u> for $ 19.95 (COCA)
(78)   Amazon <u>sells</u> the book for $ 19.95

As regards their signifieds, there does not seem to be a difference: both the *sells* of (77) and the *sells* of (78) mean 'sells'. However, while the *sells* of (77) is a **transitive verb**, which is accompanied by the object NP *the book* (cf. *The book is sold by Amazon for $ 19.95*), the *sells* of (78) is an **intransitive verb**, which is followed by the complement PP *for $ 19.95*. Recall that complements (but not objects) can be expressed by adjectives. For example, *The book sells well*.

In summary, the *sells* of (77) and the *sells* of (78) have the same meaning but different syntactics. Accordingly, these are not lexes realizing two different lexemes but allolexes of the same lexeme which, like the *early* of (75) and the *early* of (76), occur in complementary distribution.

Thanks to the OED, we know that it is the adverbial *early* which had appeared before the adjectival *early* and it is the transitive *sell* which had appeared before the intransitive *sell*. Accordingly, we can argue that adverbial *early* → adjectival *early* and transitive *sell* → intransitive *sell* are instances of lex-forming syntactics' change.

4.3.8 Lex-forming orthographic modification

Lex-forming orthographic modification is the creation of an orthographically different output allolex which has the same pronunciation as a corresponding input lex. For example, there is only an orthographic difference between the input lex *gangster* and the output allolex *gangsta*: both have the pronunciation /ˈgaŋstə/. Similarly, there is only an orthographic but not a phonetic difference between *through* and *thru*, *U.S.A.* and *USA*, *you* and *u* (as often used on the Internet), etc.

4.4 Exercises

1. Make sure you can explain each of the key terms printed in boldface (ideally, using your own examples).

2. Which of the following statements are true?

a) The terms 'lexeme-formation' and 'lex-formation' are synonyms.
b) Lexical gap is one of the reasons for lexeme-formation.
c) Nonce-formations always develop into fully-institutionalized lexemes.
d) Established lexemes have hypostatized signifieds.

Word-formation: basic issues

e) Neologisms can often be found among hapax legomena.
f) Morphological conversion is an instance of semantic change.
g) Compounding always produces allolexes of already existing lexemes.
h) New lexemes can be borrowed from other languages.
i) There are both lex-forming and lexeme-building apophonies.
j) Alphabetisms are pronounced letter by letter.

3. State which mechanisms gave rise to the following neologisms:

a) *cheapuccino* 'an inexpensive, low-quality cappuccino [...]' (Word Spy)
b) *pumpkineer* 'a person who grows giant pumpkins, particularly ones meant to be entered in pumpkin-weighing contests' (Word Spy)
c) *to Alt-Tab* 'to switch from one running computer program to another' (Word Spy)
d) *to pope* 'to convert to Catholicism' (Word Spy)
e) *9/11* 'September 11, 2001, the date of the terrorist attacks on the World Trade Center in New York and the Pentagon in Washington' (Word Spy)
f) *thumbo* 'an error made while using the thumbs to type [...]' (Word Spy)
g) *globality* 'a worldwide, interconnected economy that ignores national boundaries' (Word Spy)
h) *mission from God* 'a crucially important task that must not fail [...]' (Word Spy)
i) *soft power* 'power based on intangible or indirect influences such as culture, values, and ideology' (Word Spy)
j) *to elder* 'to share wisdom and knowledge with people who are younger than oneself' (Word Spy)

4. State which lex-forming mechanisms gave rise to the following signifiers.

a) *feck-all* (= *fuck-all*) 'nothing at all' (OED)
b) *NATO*
c) *ad*
d) *painfully*
e) *Jeff*
f) *almost* of *his almost victory*
g) *Biddy* (← *Bridgit*)
h) *math*
i) *UK*
j) *bobby* 'a slang nickname for a policeman' (OED)

5. Analyze the synchronic productivity of the following derivational suffixes using Word Spy, i.e. determine whether the database lists lexemes that have been recently created with the help of the following affixes.

a) *-er* (e.g. *teacher*)
b) *de-* (e.g. *to decentralize*)
c) *-th* (e.g. *length*)
d) *-ling* (e.g. *wolfling*)
e) *-fold* (e.g. *twofold*)
f) *neo-* (e.g. *neoconservative*)
g) *-ation* (e.g. *formation*)
h) *multi-* (e.g. *multilateral*)
i) *-eer* (e.g. *auctioneer*)
j) *in-* (e.g. *inadequate*)

4.5 Further reading

The starting point of this chapter was the classification of word-formation into lexeme- and lex-formation. This may seem a terminological innovation, but do observe that similar ideas were already expressed by Marchand (1969), who argued that mechanisms like clipping should be treated differently from compounding and derivation.

The idea that the established term 'word-formation' should be replaced by 'lexeme-formation' is also expressed by e.g. Lipka (2002: 92). Proceeding from this claim, Chapter 3 of his textbook on *English Lexicology* (Lipka 2002) provides a fairly similar classification of lexeme-building mechanisms into both formal and semantic mechanisms.

Section 4.2.6 *The establishment of new lexemes* was based on the article Schmid (2008). For a discussion of the concepts 'institutionalization' and 'lexicalization', see also the article Hohenhaus (2005).

Productivity of various lexeme-building mechanisms has been the topic of many recent studies. For example, Mühleisen (2010) deals with the productivity of the suffix *-ee* (e.g. *employee*). For an overview of how the term 'productivity' has been dealt with in lexical morphology, see the article Bauer (2005).

5 Lexeme-building mechanisms

In this chapter we will enlarge on all lexeme-building mechanisms that were briefly introduced in 4.2. Each section of this chapter begins with a discussion of the most important theoretical issues relevant for a particular lexeme-building mechanism (e.g. the directionality of conversion, differences between idiomatic adjective + noun compounds and idiomatic adjective + noun phrases) and then proceeds to the analysis of that mechanism's synchronic productivity, i.e. the question of whether the mechanism under analysis has recently been used for creating new lexemes. Special emphasis is laid on the methodological issue of how (i.e. with the help of which tools) students of English morphology themselves can determine whether a particular mechanism is still productive in Present-day English.

5.1 Semantic change

As was pointed out in 4.2.1, purely semantic mechanisms can be subsumed under the term 'semantic change'. For example, the lexeme STEALER$_1$ 'one who steals; a thief' can be said to have undergone semantic change in that it gave rise to the quasi-idiom STEALER$_2$ 'one who steals something specified'. That is, without any overt changes, the signifier *stealer* came to be associated with a different signified, thereby giving rise to a new lexeme. A similar example is *boyfriend*. As we established in 4.2.4, the fully-idiomatic lexeme BOYFRIEND$_2$ 'a male of almost any age over puberty with whom a woman has a non-marital sexual relationship' is not a product of compounding of the input roots *boy* and *friend* but of semantic change of the already existing lexeme BOYFRIEND$_1$ 'a friend who is a boy, a boyhood friend'.

What is shared by the output lexemes STEALER$_2$ and BOYFRIEND$_2$ is that their input lexemes STEALER$_1$ and BOYFRIEND$_1$ are not institutionalized lexemes of Present-day English. That is, the signifier *stealer* is now associated only with the quasi-idiomatic signified 'one who steals something specified' and the signifier *boyfriend* is now chiefly associated with the fully-idiomatic signified 'a male sexual partner'. This, however, is not the default outcome of semantic change. In the majority of cases, input and output lexemes coexist within the same vocable. For example, the output lexeme TRUE$_2$ 'not false' coexists with the input lexeme TRUE$_1$ 'faithful'; the output lexeme HAVE$_3$ 'to experience sexual relations' coexists with the input lexeme HAVE$_2$ 'to experience'; the output lexeme TO WIFE 'to downplay a woman's career accomplishments in favor of her abilities as wife and mother' coexists with the input lexeme A WIFE 'a female spouse'; etc.

5.1.1 Mechanisms of semantic change

The two mechanisms of semantic change are **metonymy** and **metaphor**. The former produces output lexemes whose signifieds are more or less objectively connected to those of their input lexemes. For example, the output lexeme BOYFRIEND$_2$ is a product of a metonymic change: as was observed in 1.2.2, boyfriends are often perceived by their girlfriends as friends. Similarly, the lexeme STEALER$_2$ can be regarded as a product of a metonymic change of STEALER$_1$: both the former and the latter share the meaning 'one who steals'.

By contrast, in the case of a metaphoric change, no such connection can be established. Consider, for instance, the lexemes FIREWALL$_1$ 'a wall that prevents the spread of fire in a building' and FIREWALL$_2$ 'a computer program that protects computers from viruses and especially hackers' attacks'. On the one hand, both FIREWALL$_1$ and FIREWALL$_2$ are protection devices: FIREWALLS$_1$ protect houses from fire and FIREWALLS$_2$ protect computers from outside intruders. On the other hand, FIREWALLS$_2$ are not walls that prevent the spread of fire. First of all, a computer firewall is not 'one of the sides of a room or building connecting floor and ceiling or foundation and roof' (MWO), which the signifier *wall* literally means. Second, computer firewalls have nothing to do with fire, for neither hackers' attacks nor viruses set computers on fire. Accordingly, the output lexeme FIREWALL$_2$ can be regarded as a product of a metaphoric semantic change of the input lexeme FIREWALL$_1$: there exists a **perceived similarity** between the senses 'a wall that prevents the spread of fire' and 'a computer program that prevents hackers' attacks' rather than an objective connection between building firewalls and their counterparts in the digital world.

A very similar example is BOOKMARK$_2$ 'a menu entry or icon on a computer that is usually created by the user and that serves as a shortcut to a previously viewed location' (MWO). As in the case of FIREWALL$_2$, BOOKMARKS$_2$ never qualify as bookmarks in the literal meaning of this word. That is, BOOKMARKS$_2$ are never 'markers for finding a place in a book' (MWO). At the same time, there is an obvious similarity between the two lexemes: both BOOKMARKS$_2$ and BOOKMARKS$_1$ can be seen as devices that enable us to easily find previously seen Web pages / places in a book that we have found particularly interesting. Accordingly, the output lexeme BOOKMARK$_2$ can also be regarded as a product of a metaphoric semantic change of the input lexeme BOOKMARK$_1$.

5.1.2 Types of metonymies

Following Kövecses and Radden (1998), metonymies can be classified into:

1. **part-for-whole metonymies**

2. **whole-for-part metonymies**
3. **part-for-part metonymies**

As an illustration of a part-for-whole metonymy, let us consider the meaning of *speaks* in (79).

(79)   Mary *speaks* Spanish (from Kövecses and Radden 1998: 52)

As Kövecses and Radden point out, the *speaks* of (79) does not only refer to the ability 'to utter words or articulate sounds with the ordinary voice' (MWO), which the signifier *speak* literally means, but to the entire linguistic proficiency, which includes not only the ability to speak but also the ability to comprehend other people's speech. Accordingly, the *speaks* of (79) can be seen as a product of a part-for-whole metonymic change: the signified of the input lexeme SPEAK$_1$ 'to utter words or articulate sounds with the ordinary voice' is part of the signified of the output lexeme SPEAK$_2$ 'to utter words as well as to comprehend words uttered by other speakers'. Part-for-whole metonymies are sometimes referred to as instances of **semantic widening**. That is, the signifier *speak* can be said to have undergone semantic widening in that it can now be used not only in connection with speaking but also in connection with the entire linguistic proficiency, of which speaking is a part.

The reverse of part-for-whole metonymies are whole-for-part metonymies. These produce output lexemes whose signifieds are narrower than those of corresponding input lexemes. In the previous chapters, we already discussed a number of output lexemes whose signifieds can be seen as products of whole-for-part metonymies. For example, the output lexeme HAVE$_3$ 'to experience sexual relations' (e.g. *He had many women*) has a narrower signified than the input lexeme HAVE$_2$ 'to experience' (e.g. *He had a good time*): experiencing sexual relations is an instance of experiencing. Similarly, the output lexeme TRUE$_2$ 'not false / faithful to the truth' has a narrower signified than the input lexeme TRUE$_1$ 'faithful': being faithful to the truth is an instance of being faithful. Both *have* and *true* can thus be said to have undergone **semantic narrowing**. (This term was already introduced in 2.2.1.)

Finally, part-for-part metonymies are metonymies that cannot be regarded as either part-for-whole or whole-for-part metonymies. Consider, for instance, the semi-idiomatic phrase *health insurance*. It is a well-known fact that a health insurance is not really a health but illness insurance (Holder 2008: 211), i.e. an insurance that covers medical costs when the insured falls ill. We do not really need a health insurance as long as we are healthy. Similarly, the UK *Department of Health* is in reality an illness department, i.e. a department that is concerned with illnesses rather than with health. Given these uses of *health*, we can argue that in English (and in many other languages) there exists the fully-idiomatic lexeme HEALTH₂ 'illnesses / pertaining to illnesses'. The semantic development undergone by *health* cannot be regarded as either a part-for-whole or a whole-for-part metonymy. An illness is neither a part nor an instance of health and health is neither a part nor an instance of an illness. Instead, we can regard the signifieds 'health' and 'illnesses' as different instances of the concept THE GENERAL CONDITION OR STATE OF THE HUMAN BODY. That is, our bodies can either be healthy or ill. Accordingly, in the case of the semi-idioms *health insurance* and *department of health*, one of the instances of the concept THE GENERAL STATE OF THE HUMAN BODY – HEALTH – stands for another instance of the same concept: ILLNESS. The metonymy HEALTH-FOR-ILLNESSES is thus a part-for-part metonymy. (We use this metonymy because of the taboo of fear or to be more precise, because of our being superstitious (Moskvin 2010: 102-106; Holder 2008: 211). That is, we are still afraid of a number of illnesses and therefore replace the taboo-marked subject ILLNESS by its converse HEALTH.)

5.1.3 Types of metaphors

Instances of a metaphoric semantic change can be classified into those that involve an **experiential similarity** or an **experiential co-occurrence** of the signifieds of their input and output lexemes (Lakoff and Johnson 1980: 154-155). With regard to the former, consider again the lexeme pairs FIREWALL₁–FIREWALL₂ and BOOKMARK₁–BOOKMARK₂. As was noted in 5.1.1, computer firewalls are similar to building firewalls in that both the former and the latter serve as protection devices: computer firewalls protect computers from outside intruders and building firewalls protect buildings from the spread of fire. Likewise, Internet bookmarks are similar to traditional book bookmarks in that both enable us to easily find the information that we find particularly interesting and important.

As far as experiential co-occurrence is concerned, consider the metaphoric use of *big* in (80).

(80)   *Tomorrow is a big day for me* (from Lakoff and Johnson 1999: 50)

The *big* of (80) obviously does not mean 'big' but 'of great importance or significance' (MWO): a big day is a very important day. In contrast to the lexemes FIREWALL₂ and BOOKMARK₂, the output lexeme BIG₂ 'of great importance' does not exhibit an experiential similarity to the lexeme BIG₁ 'big': important days are never literally bigger than unimportant days. What accounts for the metaphoric use of *big* is the existence of things that are simultaneously big and important. As Lakoff and Johnson (1999: 50) conjecture, as children, we often discover that big things such as e.g. our parents are not only big but also important. This experiential co-occurrence of objects that simultaneously possess the characteristics of being big and being important could have given rise to the metaphoric lexeme BIG$_2$ 'of great importance'.

5.1.4 Morphological conversion

Morphological conversion is a type of semantic change that produces output lexemes whose lexes are members of different word classes than the lexes of their input lexemes. For example, as we established in 4.2.1, while the input lexemes A WIFE and THOMAS are realized by the nominal lexes *a wife* and *Thomas*, the output lexemes TO WIFE and TO THOMAS are realized by the verbal lexes *to wife* and *to Thomas*. The explanation for this is that the signifieds of the converted output lexemes are headed by verbal semantic components. That is, the quasi-idiomatic signified of the output lexeme TO WIFE can be represented as the VP *to downplay a woman's career accomplishments in favor of her abilities as wife and mother*, whose head is the verb *downplay*. Similarly, the fully-idiomatic signified of the lexeme TO THOMAS is representable as the VP *to accuse a person of sexual harassment*, whose head is the verb *accuse*. This accounts for the fact that both the output lexes *to wife* and *to Thomas* are verbs, whereas the input lexes *a wife* and *Thomas*, which are headed by the nominal meanings 'a wife' and 'Thomas', are nouns. The syntactics' change undergone by the input lexes *a wife* and *Thomas* can thus be regarded as a peripheral consequence of the semantic change undergone by the input signifieds 'a wife' and 'Thomas'.

Converted lexemes are usually products of a metonymic semantic change. For example, the verbal lexemes TO WIFE and TO THOMAS are products of a part-for-whole metonymic modification of the corresponding nominal lexemes A WIFE and THOMAS. Thus the nominal concept of a wife is part of the verbal concept of downplaying a woman's career accomplishments in favor of her abilities as wife and mother. Similarly, the concept of the person called *Clarence Thomas* was originally part of the concept of accusing a person of sexual harassment. At the same time, notice that metaphoric conversion does exist as well. Consider, for example, the converted verb *to ape* 'to imitate, mimic' (OED). In contrast to both the output signifieds 'to wife' and 'to Thomas', the output signified 'to ape' is a product of metaphorization of the input signified 'an ape': there is only a perceived similarity between an ape and a person who apes.

As regards the semantic outcome of conversion, most converted lexemes can be analyzed as quasi-idioms in relation to their corresponding input lexemes, i.e. the signifieds of the former contain the signifieds inherent in the latter plus some additional idiomatic signifieds. For example, the output lexeme TO WIFE is a quasi-idiom in relation to the input lexeme A WIFE: the signified 'to wife' contains the signified 'a wife', inherent in the component *wife*, plus the idiomatic signified 'to downplay a woman's career accomplishments in favor of her abilities as mother'. The same applies to the converted lexemes TO BACKGROUND 'to surreptitiously perform a task in the background while one's attention is supposed to be on another task', TO OFFICE 'to perform office-related tasks, such as photocopying and faxing', and TO PIE 'to hit a person, particularly a political or business leader, in the face with a pie', which were mentioned in 4.2.8.

Fully-idiomatic converted lexemes are usually instances of a metaphoric semantic change like TO APE as well as metonymies like TO THOMAS whose input lexemes are the names of people, places, institutions, etc. who / which played some role in what later came to be referred to by means of their converted output lexemes. For example, the converted lexeme TO DELL 'to beat a competitor by eliminating the middleman and selling directly to customers' (Word Spy) is, like TO THOMAS, a full-idiom: its signified does not contain the signified of the input lexeme DELL. The latter only motivates the former: according to Word Spy, TO DELL

> comes from Dell Computers, which used direct sales to become one of the top computer manufacturers in the world, with current sales of US$18 billion. (Word Spy)

From a formal point of view, converted lexemes are usually classified with regard to the **direction of conversion**. For example:

- The output verbal lexemes TO APE, TO BACKGROUND, TO DELL, TO OFFICE, TO PIE, TO THOMAS, and TO WIFE, whose input lexemes are the nominal lexemes AN APE, A BACKGROUND, DELL, AN OFFICE, A PIE, THOMAS, and A WIFE, are instances of **noun → verb conversion**.

- The output verbal lexeme TO ELDER 'to share wisdom and knowledge with people who are younger than oneself' (Word Spy), whose input lexeme is the adjectival lexeme ELDER, is an instance of **adjective → verb conversion**.

- The output verbal lexeme TO DOWN 'to bring, put, throw, or knock down' (OED), whose input lexeme is the adverbial lexeme DOWN, is an instance of **adverb → verb conversion**.

- The output nominal lexeme A SHOUT-OUT 'a mention, acknowledgement, or greeting, especially one made over the radio or during a live performance; a namecheck' (OED), whose input lexeme is the verbal lexeme TO SHOUT OUT, is an instance of **verb → noun conversion**.

- The output nominal lexeme AN EX 'one who formerly occupied the position or office denoted by the context; specially a former husband or wife' (OED), which goes back to the prefix *ex-* of *ex-husband* and *ex-wife*, is an instance of **affix → noun conversion**.

Of particular theoretical interest are **adjective → noun conversion** and **noun → adjective conversion**. As regards the former, let us consider the nominal lexeme A GOLD 'a medal awarded as the first prize in a competition: a gold medal' (MWO). At first glance, it seems that this lexeme is a product of metonymy-based quasi-idiomatization of the input adjectival lexeme GOLD. That is, the output nominal signified 'a gold medal' contains the adjectival signified 'gold', inherent in the input adjective *gold*, plus the idiomatic signified 'medal'. The same seems to be true of the nominal lexeme A FINAL 'the last examination in a course' (MWO), which is often realized by the plural lex *finals*: its signified contains the adjectival signified 'final', inherent in the input adjective *final*, plus the idiomatic signified 'exam in a course'. However, in spite of these facts, neither A GOLD nor A FINAL can be regarded as products of adjective → noun conversion. In the case of these lexemes, we are dealing not with a lexeme-building conversion but with a lex-forming clipping of the semi-idiomatic NP *a gold medal* and the quasi-idiomatic NP *a final examination*. That is, *a gold* ← *a gold medal* and *a final* ← *a final examination*. This analysis is true not only from a diachronic but also from a synchronic point of view. That is, present-day English speakers still analyze *a gold* and *a final* as shorter versions of the input NPs *a gold medal* and *a final examination*. Accordingly, cases like *a final* ← *final* and *a gold* ← *gold* must be regarded as **pseudo-conversion**.

As regards noun → adjective conversion, consider so-called ***stone wall-combinations***. These are NPs in which the modifier position is filled by a noun: e.g. *a phone call, a police officer, a car accident, a movie star,* etc. In these NPs, the modifiers *phone, police, car,* and *movie* fill the syntactic slot that is typically filled by adjectives. That is, adjectives typically function as modifiers of head nouns in NPs. For example, *a recent call, a kind officer, a horrible accident, a popular star*. Given this fact, a question arises as to whether nouns like *phone, police, car,* and *movie* that fill the modifier position in NPs are not really nouns but adjectives. If so, we will be justified in concluding that in English adjectival lexemes can come into existence via noun → adjective conversion.

To answer this question, let us recall in which respects adjectives are syntactically different from members of other word classes. As we observed in 2.2.3, adjectives function as either modifiers of head nouns in NPs (e.g. *a horrible accident*) or as complements of predicator verbs in clauses like *The accident was horrible*. Accordingly, to determine whether *phone, police, car,* and *movie* of the NPs *a phone call, a police officer, a car accident,* and *a movie star* are indeed adjectives, we must try to place them in complement positions. Consider, for example, (81), (82), (83), and (84).

(81) *This call is phone** ('this call is being made with the help of a telephone')
(82) *This officer is police** ('this officer is a policeman')
(83) *This accident is car** ('this accident involves a car')
(84) *This star is movie** ('this person has become famous because of the movies in which he or she played the main roles')

Given the obvious ungrammaticality of these invented sentences, we can conclude that the modifiers *phone, police, car,* and *movie* of the NPs *a phone call, a police officer, a car accident,* and *a movie star* are not converted adjectives but nouns functioning as modifiers of the head nouns *call, officer, accident,* and *star*.

However, it is important to emphasize that this does not apply to all *stone wall*-combinations. Consider, for example, the NP *a stone wall* (which gave the name to the *stone wall*-problem). According to MWO, in Present-day English there is the adjectival lexeme STONE$_2$ meaning 'of, relating to, or made of stone'. That is, we can say *This wall is stone* referring to a wall made of stone. Usually, the adjective *stone* is used metaphorically. For example, Cindy Lauper's 1992 song *The World is Stone* deals with the cruelty and the indifference of the contemporary world. This metaphorical use is, however, possible because of the literal adjectival meaning 'made of stone'. Accordingly, the output adjective *stone* of the NP *a stone wall* can be regarded as a product of noun → adjective conversion of the input noun *stone*. We are justified in claiming this not only because *stone* can fill both the modifier and the complement position but also because the output adjectival lexeme is a quasi-idiom in relation to the input nominal lexeme: the adjectival signified 'of, relating to, or made of stone' does not only contain the signified 'stone', inherent in the component *stone*, but also the idiomatic signified 'of, relating to, or made of'. The adjective *stone* is thus clearly a lex which realizes a new lexeme.

Finally, let us discuss the practical problem of determining the direction of conversion. As an illustrative example, let us consider the converted verbal lexeme TO KNIFE 'to use a knife to; to cut, strike, or stab with a knife' (OED). How do we actually know that it is the nominal lexeme A KNIFE which served as an input lexeme for TO KNIFE, not vice versa? As we will see below, this question can be approached from both a diachronic and a synchronic perspective.

As regards the former, we know (thanks to the OED) that the verb *to knife* appeared in the English language much later than the noun *a knife*. While the latter has been recorded since 1100, the former has been used only since circa

1885 (OED). However, this fact alone does not suffice to conclude that it was the nominal lexeme A KNIFE which served as an input lexeme for the verbal lexeme TO KNIFE. The main reason why the latter can be regarded as an instance of conversion of the former is that the verbal meaning 'to use a knife to; to cut, strike, or stab with a knife' is semantically more complex than the nominal meaning 'a knife'. The verbal meaning is a quasi-idiom in relation to the nominal meaning: the signified 'to knife' contains the signified 'a knife', inherent in the component *knife*, plus the idiomatic signified 'to cut, strike, or stab'. Accordingly, given the fact that *to knife* appeared later than *a knife* and given that the signified 'to knife' is a quasi-idiom in relation to the signified 'a knife', we are justified in claiming that it is the nominal lexeme A KNIFE which served as an input lexeme for the verbal lexeme TO KNIFE, not vice versa.

Similarly, thanks to the OED, we know that the verb *to ape* appeared much later (1634) than the noun *an ape* (700). However, as in the case of *to knife*, this fact alone does not suffice to regard the verbal lexeme TO APE as a product of conversion of the nominal lexeme AN APE. The main reason why we can do so is that the output verbal signified 'to mimic' is a product of metaphorization of the input nominal signified 'an ape': as stated above, there is a perceived similarity between an ape (animal) and a person who apes.

Sometimes the diachronic analysis is not as straightforward as in the case of TO KNIFE and TO APE. Consider, for example, the verbal vocable **TO DOCTOR**. According to the OED, this vocable consists of the following six lexemes.

- TO DOCTOR$_1$ 'to confer the degree or title of Doctor upon; to make a Doctor' (OED / 1599)

- TO DOCTOR$_2$ 'to treat, as a doctor or physician; to administer medicine or medical treatment' (OED / 1740)

- TO DOCTOR$_3$ 'to repair, patch up, set to rights' (OED / 1829)

- TO DOCTOR$_4$ 'to castrate an animal'(OED / 1902)

- TO DOCTOR$_5$ 'to treat so as to alter the appearance, flavor, or character of; to disguise, falsify, tamper with, adulterate, sophisticate' (OED / 1777)

- TO DOCTOR$_6$ 'to take medicine, undergo medical treatment' (OED / 1865)

Each of these lexemes appeared after the noun *doctor*, which has been documented in English since the 14th century. The OED lists a number of nominal signifieds that *doctor* could express at that time. For example:

- 'a teacher, instructor; one who gives instruction in some branch of knowledge, or inculcates opinions or principles' (OED / 1387)

- 'one who, by reason of his skill in any branch of knowledge, is competent to teach it, or whose attainments entitle him to express an authoritative opinion; an eminently learned man' (OED / 1400)

- 'a person who, in any faculty or branch of learning, has attained to the highest degree conferred by a University; a title originally implying competency to teach such subject or subjects, but now merely regarded as a certificate of the highest proficiency therein' (OED / 1377)

- 'a doctor of medicine; in popular current use, applied to any medical practitioner. Also: (amongst indigenous peoples) a traditional healer or diviner, esp. one dealing with afflictions thought to be caused by spirit possession or witchcraft' (OED / 1377)

However, despite the fact that the noun *doctor* appeared before the verb *to doctor*, we cannot claim that each of the six verbal TO DOCTOR lexemes are instances of noun → verb conversion. Let us begin with the lexeme TO DOCTOR$_1$ 'to confer the degree or title of Doctor upon; to make a Doctor' (OED / 1599). This verbal lexeme is a quasi-idiom in relation to the nominal lexeme DOCTOR$_1$ 'a person who has earned the degree of a doctor' (OED / 1377): the signified of TO DOCTOR$_1$ contains the signified 'a doctor', inherent in the component *doctor*, plus the idiomatic signified 'to confer this degree upon somebody'. Accordingly, given the quasi-idiomatic character of the relation between the two lexemes under consideration and given that the verbal lexeme appeared after the corresponding nominal lexeme, we are justified in concluding that TO DOCTOR$_1$ is an instance of noun → verb conversion of DOCTOR$_1$.

A similar analysis applies to the lexeme TO DOCTOR$_2$ 'to treat, as a doctor or physician; to administer medicine or medical treatment' (OED / 1740). This lexeme is a quasi-idiom in relation to the nominal lexeme DOCTOR$_2$ 'a medical practitioner' (OED / 1377). The signified of TO DOCTOR$_2$ contains the signified 'a doctor', inherent in the component *doctor*, plus the idiomatic signified 'to do the job of'. Accordingly, given this fact and given that the verbal lexeme appeared later than the corresponding nominal lexeme, we are justified in concluding that TO DOCTOR$_2$ is also a product of noun → verb conversion of the nominal lexeme DOCTOR$_2$.

But what about the lexeme TO DOCTOR$_3$ 'to repair, patch up, set to rights' (OED / 1829)? In contrast to the lexemes discussed above, it appears that TO DOCTOR$_3$ is not an instance of noun → verb conversion of any of the nominal DOCTOR lexemes but a product of part-for-whole metonymization of the verbal lexeme TO DOCTOR₂ 'to treat, as a doctor or physician; to administer medicine or medical treatment' (OED / 1740). Thus the signified 'to treat as a doctor or physician' seems to be an instance of the signified 'to repair'. That is, giving a medical treatment to somebody often results in the repairing of that person's health.

Similarly, the lexemes TO DOCTOR₄ 'to castrate an animal' (OED / 1902) and especially TO DOCTOR₆ 'to take medicine' (OED / 1865) seem to be products of a metonymic semantic change of the already existing verbal lexeme TO DOCTOR₂: castrating an animal can be seen as an instance of giving medical treatment to that animal and taking medicine can be seen as giving medical treatment to oneself. Hence the lexemes TO DOCTOR₄ and TO DOCTOR₆ can be regarded as products of whole-for-part metonymization of the lexeme TO DOCTOR₂.

Finally, let us consider the lexeme TO DOCTOR₅ 'to falsify' (OED / 1777). The etymology of this lexeme is not entirely clear. On the one hand, it may have been the case that the signifier *to doctor* acquired this meaning because of some people's belief that doctors (be it holders of doctoral degrees or practitioners of medicine) often falsify things. In this case, TO DOCTOR₅ can be regarded as an instance of noun → verb conversion of any of the nominal DOCTOR lexemes. On the other hand, however, it can be conjectured that any instance of doctoring (be it conferring the degree of a doctor upon somebody or giving medical treatment to somebody) involves falsifications. In this case, TO DOCTOR₅ can be regarded as a product of metonymy-based semantic change of one of the previously formed TO DOCTOR lexemes.

In summary, of the six verbal TO DOCTOR lexemes, only two can be considered products of morphological conversion of two different nominal DOCTOR lexemes, whereas the other four lexemes are instances of a metonymic semantic change (not involving a change of word class) of the previously formed verbal TO DOCTOR lexemes.

As a rule, the diachronic history of the members of a pair like A KNIFE–TO KNIFE confirms the intuitions of present-day speakers who do not have the diachronic memory (i.e. do not remember the true etymologies of the lexemes in question). Indeed, in Present-day English we still have the noun–verb pair A KNIFE–TO KNIFE, in which the verbal lexeme can be analyzed as a quasi-idiom in relation to the corresponding nominal lexeme: the signified 'to knife' contains the signified 'a knife', inherent in the component *knife*, plus the idiomatic signified 'to use a knife in a particular way: to cut, strike, or stab with a knife'. Accordingly, even without looking up the diachronic history of the lexemes A KNIFE and TO KNIFE in a dictionary like the OED, we can claim that the verbal lexeme TO KNIFE is, from a synchronic point of a view, a product of conversion of the nominal lexeme A KNIFE. Likewise, we can claim that the verbal lexeme TO APE is, from a synchronic point of a view, a product of conversion of the nominal lexeme AN APE: the verbal signified 'to ape' can still be analyzed as an instance of metaphorization of the nominal signified 'an ape'.

However, as will be shown below, in a number of cases the diachronic history is at odds with present-day speakers' intuitions. Consider, for example, the lexemes AN ANSWER and TO ANSWER. From a synchronic point of view, we seem to be dealing with a conversion pair one of whose members can be analyzed as a quasi-idiom in relation to the other member: *an answer* seems to mean 'the product of the process of answering'. However, the diachronic history of the lexemes under analysis does not corroborate this conjecture. According to the OED, the nominal lexeme AN ANSWER originally (i.e. in the Old English period) was realized by the lex *andswaru*, whereas the verbal lexeme TO ANSWER was originally realized by the lex *andswarian*. In other words, the lexemes AN ANSWER and TO ANSWER originally had different lexes: *andswaru* and *andswarian* had the same root *andswar* but different derivational affixes: -*u* and -*ian*. (With the course of time, *andswaru* and *andswarian* 'got rid' of these suffixes, so that in Present-day English both the nominal lexeme AN ANSWER and the verbal lexeme TO ANSWER are realized by the same lex *answer*.) A very similar case is the lexeme pair LOVE–TO LOVE, whose members were originally realized by the non-identical lexes *lufu* (*love*) and *lufian* (*to love*). Again, as in the case of AN ANSWER and TO ANSWER, the pair LOVE–TO LOVE can be regarded as a conversion pair only from a synchronic but not from a diachronic point of view. (From a diachronic perspective, the verbal lexes *andswarian* and *lufian* can perhaps be regarded as output lexes which came into existence via affixation of the input nominal lexes *andswaru* and *lufu* by means of the derivational suffix -*ian*.)

A number of studies provide additional synchronic criteria for determining the direction in conversion. For example, Ginzburg et al. (1976: 133-134) mention the **synonymic sets criterion**. As they argue, to determine the direction of conversion in the pairs A CHAT–TO CHAT and A SHOW–TO SHOW, we need to consider the near-synonymic pairs A CONVERSATION–TO CONVERSE and AN EXHIBITION–TO EXHIBIT, in which the lexes realizing the nominal lexemes can be synchronically analyzed as products of affixation of the lexes realizing the corresponding verbal lexemes: *a conversation* = *to converse* + -*ation* and *an exhibition* = *to exhibit* + -*ion*.[11] Accordingly, given the derived character of the nominal lexemes A CONVERSATION and AN EXHIBITION and given a semantic similarity between the signifieds 'conversation' / 'chat' and 'exhibition' / 'show', it can be conjectured that in the conversion pairs A CHAT–TO CHAT and A SHOW–TO SHOW, it is the nominal lexemes A CHAT and A SHOW that are products of conversion of the corresponding verbal lexemes TO CHAT and TO SHOW, not vice versa.

---

[11] Note that this analysis is not true from a diachronic point of view: According to the OED, both *conversation* and *exhibition* were borrowed into English from Old French.

Another often mentioned criterion is the **frequency criterion** (Ginzburg et al. 1976: 135-136; Plag 2003: 111). As is well-known, output lexemes that came into existence via derivation and conversion are often characterized by a lower frequency of use than corresponding input lexemes. The reason for this is that the signifieds of output lexemes (especially those that came into existence via quasi-idiomatization) are more complex than those of corresponding input lexemes. As a result of this semantic complexity, the former cannot be used as freely as the latter: the use of semantically more complex output lexemes is constrained by additional meanings which they have acquired as a result of their quasi-idiomatization. What follows from this is that the direction of conversion can be determined by comparing the frequencies of use of the members of a conversion pair. For example, if you use COCA or BYU-BNC, this can be done in the following way. Just enter the combination [chat].[v*] to the search mask and both corpora will yield all verbal occurrences of the verb *to chat*, including the wordforms *chatting, chats, chatted*. If you enter [chat].[n*], both corpora will yield all nominal occurrences of the noun *chat*, including the plural wordform *chats*. At present (i.e. as of November 28, 2011), COCA finds 3071 occurrences of the nominal lexeme A CHAT and 5201 occurrences of the verbal lexeme TO CHAT. The lower frequency of *a chat* testifies to the analysis of this noun as a product of verb → noun conversion of the verb *to chat*. As regards *a show* and *to show*, COCA yields 225869 occurrences of the verb *to show* but only 96566 occurrences of the noun *a show*. Again, this can be seen as a corroboration of the analysis of the noun *a show* as a product of conversion of the verb *to show*.

Both of these criteria are problematic. As regards the synonymic sets criterion, its major shortcoming acknowledged by Ginzburg et al. is its somewhat limited applicability: it is not always possible to find a synonymic pair like *to converse–a conversation* one of whose members can be analyzed as a product of affixation of the other member.

Similarly, the frequency criterion is of very little help when we have to deal with polysemous lexemes. As an illustrative example, let us again consider the verb *to doctor*. Using COCA, we can perhaps prove that this verb came into existence via noun → verb conversion of the noun *doctor*: while a search for [doctor].[v*] yields only 56 results, a similar search for [doctor].[n*] yields 75565 results. However, the only thing these numbers tell us is that some of the nominal DOCTOR lexemes must have given rise to some of the six verbal TO DOCTOR lexemes. This, however, is an incomplete analysis: as we have established above, of the six TO DOCTOR lexemes, no more than two can be regarded as instances of conversion.

But even if it were possible to compare the frequencies of different TO DOCTOR lexemes with those of the corresponding nominal lexemes, the frequency criterion would still be of fairly marginal significance. Thus the only reason why the noun *chat* can be regarded as a product of conversion of the verb *to chat* is that the signified of the nominal lexeme represents a quasi-idiom in relation to the signified of the corresponding verbal lexeme: a chat is the product of chatting. If you do not analyze the signified 'a chat' as being semantically more complex than the signified 'to chat', you cannot claim that the former came into existence via conversion of the latter even if you can show that *a chat* has a lower frequency of use than *to chat*.

To conclude: the most important criterion for determining the direction of conversion from both diachronic and synchronic perspectives is the **semantic criterion**. In the case of converted lexemes that came into existence via metonymy-based quasi-idiomatization, the signified of a converted output lexeme is semantically more complex than the signified of a corresponding input lexeme. Similarly, in the case of converted lexemes which came into existence via metaphor-based full-idiomatization, the signified of an output lexeme can be analyzed as a product of metaphorization of the signified of a corresponding input lexeme. No other criterion justifies the treatment of the lexeme under analysis as an instance of morphological conversion of some other lexeme in question.

5.1.5 Productivity

Semantic change is the most productive lexeme-building mechanism.[12] People are constantly changing the meanings of already existing lexemes in an attempt to produce an impression on other people (i.e. being motivated by the wish to achieve success). Compared to other lexeme-building mechanisms, semantic modification requires the least effort on the part of a language user, i.e. he or she only has to modify the signified without modifying or inventing the signifier.

The claim that semantic change is more productive than other mechanisms can be corroborated by the structure of English vocables. Most of them consist of multiple polysemous lexemes. Recall that, for example, the vocable TRUE consists of the polysemous lexemes $TRUE_1$ 'faithful', $TRUE_2$ 'not false', $TRUE_3$ 'properly so called', $TRUE_4$ 'legitimate, rightful', $TRUE_5$ 'narrow, strict', etc. One-lexeme vocables can be found as well, but they are relatively rare. The reason for this is the above mentioned fact that language users, motivated by the wish to produce an impression on other people, are constantly changing the signifieds of already existing lexemes, thereby creating new lexemes.

As regards morphological conversion, it is the noun → verb pattern, exemplified by AN APE → TO APE, A BACKGROUND → TO BACKGROUND, A KNIFE → TO KNIFE, etc., which is more productive than all other patterns named in

---

[12] I owe the understanding of this to Dieter Stein.

5.1.4. As, for example, Ginzburg et al. (1976: 138) point out, "the possibility for the verbs to be formed from nouns through conversion seems to be illimitable". The explanation for this is the fact that in Present-day English there are only two derivational suffixes that can be used for deriving verbs from nouns. These are the suffix *-ify* of e.g. *personify* and the suffix *-ize* of e.g. *actualize*. Both suffixes are still productive nowadays; however, as Marchand (1969: 364) notes, they have "restricted range of derivative force": "*-ify* is learned while *-ize* is chiefly technical". Indeed, of 55 verbal lexemes that appeared in the English language between 1990 and 2011, we find only one instance of *-ify* affixation – TO MATTIFY 'of a cosmetic: to produce a matt appearance; to reduce the shiny appearance of the skin, esp. on the face' (OED / 1997)[13] – and only four instances of *-ize* affixation:

- TO CLINTONIZE 'to modify in accordance with or as a result of the policies of President Clinton' (OED / 1992)

- TO FERBERIZE 'to use the Ferber method or a similar technique to train (a child) to fall asleep independently' (OED / 1990)

- TO MODULIZE 'to render modular; to modularize' (OED / 1992)

- TO TALIBANIZE 'to treat in a manner associated with the Taliban [...]' (OED / 1997)

By contrast, 29 verbal lexemes (of the 55 verbal lexemes which appeared during the last two decades) are products of noun → verb conversion. Consider, for example, the following lexemes.

- TO BITCH-SLAP 'to deliver a stinging slap to (a person), especially in order to humiliate one regarded as inferior' (OED / 1991)

- TO BOTOX 'to treat with Botox' (OED / 1994)

- TO DOTCOM 'to be overwhelmed or driven out of existence by pressure from companies doing business over the Internet' (OED / 1996)

- TO GOOGLE 'to use the Google search engine to find information on the Internet' (OED / 1999)

---

[13] Note that *to mattify* is a product of affixation of the input adjective *matt* 'of a surface, finish, etc.: without lustre, dull; unpolished' (OED) by means of the suffix *-ify*.

- TO HOT-DESK 'to practice hot-desking; to work at any of a number of different desks or workstations on a temporary, ad hoc, or part-time basis, rather than routinely occupying one desk permanently; to allocate workspace in this way' (OED / 1994)

- TO NINJA 'to act or move in a manner similar to a ninja' (OED / 1992)

- TO PEPPER-SPRAY 'to douse with pepper spray' (OED / 1993)

- TO TEXT MESSAGE 'to send or enter as a text message; to send a text message to (a person)' (OED / 1994)

- TO TWOC 'to steal (a car), especially for the purpose of joy-riding' (OED / 1992)

Noun → verb conversion can thus be regarded as the most productive verb-forming mechanism in Present-day English.

Unfortunately, the Etymology-sections of most OED entries do not contain the word *conversion*. For example, the Etymology-section of the entry for TO BITCH-SLAP, which, as stated above, came into existence via noun → verb conversion of the nominal lexeme A BITCH SLAP, contains the following information: 'Etymology: < *bitch slap* n.' (OED). This information enables us to recognize that the nominal lexeme A BITCH SLAP served as an input lexeme for the verbal lexeme TO BITCH SLAP. However, since the direction sign < also occurs in the Etymology-sections of the lexemes that came into existence via other mechanisms (e.g. the Etymology-section of the affixed verb *to modulize* contains the information 'Etymology: < *module* n. + *-ize* suffix'), this sign is not very helpful for making the OED searchable for instances of conversion. That is, there does not seem to be a way of making the OED specifically search for converted lexemes. This is why the best means to study the productivity of various conversion patterns is to ask the OED to yield all lexemes belonging to a particular word class that were formed during a particular period of time. For instance, if you want to find out whether verb → noun conversion has recently created more / less nominal lexemes than a particular noun-building affix, go to the OED Advanced Search page http://www.oed.com/advancedsearch, enter '1990-' (without quotation marks) to 'Date of entry', and choose the option 'Noun' in 'Part of speech'. The OED will then yield all nominal lexemes which appeared in the English language between 1990 and 2011. Then carefully read the Etymology-sections of all found lexemes and assign them to the categories 'affixed nominal lexemes' and 'converted nominal lexemes'.

If you are interested in other recent instances of semantic change (i.e. those not involving a change of word class), go again to the above named OED

Advanced Search page http://www.oed.com/advancedsearch, enter e.g. '2010' (without quotation marks) to the first row of the search mask from above, and choose the option 'Quotation date' (near the first row of the search mask to which you will enter '2010'). The OED will then yield all entries containing the year *2010*. One of such entries could be a vocable one of whose lexemes has recently undergone semantic change, thereby giving rise to a new lexeme which has been registered in English since the year 2010. Of course, you can refine your search in a variety of ways. For example, if you choose the option 'Noun' in 'Part of speech', the OED will yield nominal lexemes. If you choose the option 'Colloquial' in 'Usage', the OED will yield informal lexemes.

5.2 Lexeme-manufacturing

Lexeme-manufacturing or arbitrary formation is the formation of a new lexeme by means of creating an unmotivated signifier for expressing a previously unexpressed signified. For example, as we pointed out in 4.2.2, the unmotivated proprietary names *Viagra* and *Gonk* were created to express the previously unexpressed signifieds 'a new impotence drug' and 'an egg-shaped doll'.

The most important theoretical question raised by this lexeme-building mechanism is whether any combination of sounds can become a lex realizing a new lexeme. That is, for example, does it make a difference whether the creator of an arbitrary lex chooses the signifier *plafe* or *lpafe*? Or more generally: which of these signifiers has more chances to become a lex of a new lexeme?

The answer to these questions is provided by the well-known **sonority sequencing generalization** (McMahon 2002: 107; Yavaş 2006: 127-151). In phonetics and phonology, the term 'sonority' refers to the loudness of a sound (compared to other sounds.) The two main factors determining the sonority of the sound are **the degree of opening of the vocal tract** and **voicing**.

As regards the former, a greater opening of the vocal tract results in a greater sonority. Thus vowels are acoustically much louder than consonants because they are articulated in such a way that the airflow coming from the lungs can freely leave the mouth cavity without encountering any obstruction. In contrast, the articulation of a consonant involves a major obstruction to the airflow. For example, in the case of the **plosive** [p], there is a complete obstruction formed by the upper and the lower lip, which are firmly pressed against each other: the airflow can leave the mouth cavity only when this obstruction is removed. (When the obstruction is removed, we hear a kind of explosion. Hence the term 'plosive'.)

The term 'voicing' refers to the vibration of the vocal folds (sometimes also referred to as **vocal cords**). When they are vibrating, a **voiced sound** is produced. Vowels are always voiced, whereas consonants can be both voiced and **voiceless**. For example, while [b] is voiced, [p] is voiceless.

The sonority sequencing generalization is concerned with the question of which combinations of sounds qualify as **permissible syllables**. A **syllable** is a phonetic unit that often consists of the following three structural components:

1. **the onset**
2. **the nucleus**
3. **the coda**

The nucleus is the most sonorous element of a syllable. Since vowels are more sonorous than consonants, the nucleus position is typically filled by vowels. For example, in the putative word *plafe* only the diphthong /eɪ/ can be the nucleus. The nucleus is the only obligatory element of a syllable. Thus there are syllables consisting of nuclei only. For example, /ʌɪ/ of *I*.

The onset is a consonant or a consonant cluster occurring before the nucleus. For example, the cluster [pl] is the onset of the syllable /pleɪf/.

Finally, the coda is a consonant or a consonant cluster which occurs after the nucleus. For example, the consonant [f] is the coda of the syllable /pleɪf/.

According to the sonority sequencing generalization, onsets that consist of more than one consonant are characterized by an increasing sonority. That is, the following sound has a higher sonority than the preceding one. For example, the double onset [pl] is a permissible onset because the following sound [l] is more sonorous than [p]. Both [p] and [l] involve a complete obstruction to the airflow. However, [l] is a **lateral sound**. This means that despite the obstruction formed by the raising of the tip of the tongue towards the alveolar ridge, the airflow coming from the lungs can escape from the mouth cavity along the sides of the tongue. By contrast, as stated above, in the case of the plosive [p], the airflow can leave the mouth cavity only when the obstruction formed by the lower and the upper lip is removed. In addition to this, the lateral [l] is a voiced sound, whereas the plosive [p] is a voiceless one.

That [pl] is indeed a permissible double onset in English is corroborated by the existence of numerous signifiers beginning with this cluster. For example, *plate* /pleɪt/, *plan* /plan/, *plum* /plʌm/.

However, there are no signifiers beginning with [lp]. E.g. **lpate*, **lpan*, **lpum*, etc. This consonant cluster is a non-permissible double onset because it does not fulfill the sonority sequencing generalization: the lateral [l] has a greater sonority than the plosive [p].

With regard to codas consisting of more than one consonant, the sonority sequencing generalization states that such codas must be characterized by a decreasing sonority. That is, the following consonant must have a lower sonority than the preceding one. This explains, for example, why English has the signifier *help* /hɛlp/, but not *hepl /hɛpl/. The cluster [lp] is a permissible double coda because the following plosive [p] is less sonorous than the preceding lateral [l]. By contrast, [pl] is a non-permissible double coda because it is characterized by an increasing sonority.

Taking all this into account, we can now conclude that only *plafe*, but not *lpafe*, can become an arbitrary lex of a new lexeme: in *plafe* the cluster [pl] is a permissible double onset, the diphthong /eɪ/ is a permissible nucleus, and the consonant [f] is a permissible single coda; e.g. *chief* /tʃiːf/, *roof* /ruːf/, etc.

5.2.1 Productivity

Lexeme-manufacturing requires a considerable effort on the part of a language user: he or she has to create a totally unmotivated signifier that has not been used before. Precisely because of this fact, lexeme-manufacturing is the least productive lexeme-building mechanism: as e.g. Algeo (1998: 66) points out, arbitrary formations like *plafe* "are extremely rare, if they exist at all". Indeed, according to the OED, of 587 new lexemes which appeared in the English language between 1990 and 2011, only three can be regarded as instances of lexeme-manufacturing. Apart from the already mentioned proprietary name *Viagra*, these include:

- NEVIRAPINE 'a drug structurally related to the benzodiazepines which inhibits the reverse transcriptase of HIV-1' (OED / 1991)

- ZORB 'a large, transparent, inflatable PVC ball used in the sport of zorbing, containing an inner capsule into which a participant is secured and then rolled along the ground, down hills, etc.' (OED / 1996)

Note that as in the case of *Viagra*, the OED is not entirely sure whether *nevirapine* and *Zorb* are indeed arbitrary formations. Thus *nevirapine* is, according to the OED, "apparently an arbitrary formation, probably including the elements *vir-* (in *virus* n.) and *-pine* (compare *benzodiazepine* n.)". Similarly, *Zorb* is, according to the OED, "apparently an arbitrary formation", there being a possibility that its formation was influenced by *orb* 'each of the concentric hollow spheres formerly believed to surround the earth and carry the planets [...]' (OED / circa 1449).

Earlier examples of arbitrary formations include:

- GOOP 'a stupid or fatuous person' (OED / 1900)

- TO GROK 'to understand intuitively or by empathy; to establish rapport with' (OED / 1961)

- MITHRIL 'In the fiction of J. R. R. Tolkien [...]: a rare silver-colored precious metal of great hardness and beauty' (OED / 1944)

- NERF 'a type of foam rubber used esp. in the manufacture of children's toys and sports equipment' (OED / 1970)

- SUKEBIND 'name given by Stella Gibbons [...] to an imaginary plant associated with superstition and fertility, hence used allusively with reference to intense rustic passions' (OED / 1932)

To study arbitrary formations in English, go to the above named OED Advanced Search page http://www.oed.com/advancedsearch, enter 'arbitrary formation' (without quotation marks) to the first row of the search mask from above, and choose the option 'Etymology' (near the search mask to which you will enter 'arbitrary formation'). The OED will then yield all lexemes that have ever been coined via lexeme-manufacturing. Of course, you can refine your search by entering e.g. '1900-' to 'Date of entry'. In this case, the OED will yield all arbitrary formations that appeared in the English language between the years 1900 and 2011.

5.3 Lexeme-building borrowing

Lexeme-building borrowing is the importation of a foreign language signifier that gives rise to a new lexeme. As we established in 4.2.2 and 4.2.3, there are two kinds of lexeme-building borrowing:

- the borrowing of both the signifier and the signified of a foreign language lexeme: e.g. the borrowing of *kindergarten* from German.

- the borrowing of a foreign language signifier accompanied by a semantic modification of its (foreign language) input signified: e.g. the borrowing of *wiki* from Hawaiian.

Sometimes it is argued that **loan-translations** (also known as **semantic calques**) must be regarded as a distinct type of borrowing. Consider, for example, the lexeme SUPERMAN 'an ideal superior man conceived by Nietzsche as being evolved from the normal human type; loosely, a man of extraordinary power [...]' (OED / 1903). According to the OED, the signifier *superman* is a product of affixation of *man* by means of the derivational prefix *super-*. However, what is interesting here is that this instance of affixation was inspired by the German signifier *Übermensch*, popularized by Friedrich Nietzsche's famous work *Thus spoke Zarathustra*. In other words, speakers of English did not borrow the German *Übermensch* into English. What they did was translate the German components *über* and *Mensch* as *super-* and *man*, thereby giving rise to *superman*. The English *superman* is thus a loan-translation of the German *Übermensch*.

Undeniably, the creation of the English lexeme SUPERMAN was inspired by the German lexeme ÜBERMENSCH. However, the formation of the English signifier *superman* did not involve the borrowing of any German signifier: as said above, the lex under analysis is a product of affixation of *man* by means of the prefix *super-*. Accordingly, the lexeme SUPERMAN cannot be regarded as an instance of borrowing; the lexeme under analysis is an instance of prefixation, just as the lexeme UNSACKABLE, whose creation in 1980 was not inspired by any foreign language signifier.

5.3.1 Productivity

Throughout its history, English has extensively borrowed from other languages. Below are just some of the most famous examples.

- **Latin loanwords in English:**
  altar, balsam, calculator, debit, editor, to fabricate, to generate, habitual, icon, janitor, kinesis, laboratory, magnet, to narrate, to obliterate, pagan, to quote, radiation, salient, tact, ubiquity, vacuum, xenia, zeal, etc.

- **Greek loanwords in English:**
  aerodrome, bronchotomy, calligraphy, deixis, ellipse, glaucoma, hapax legomenon, iris, laconic, macron, narcosis, oligarch, pathos, strophe, telic, xenon, etc.

- **French loanwords in English:**
  to abandon, bachelor, to commence, to damage, economy, fable, garage, harmony, to identify, jewel, kilometer, lechery, machine, naïve, obedience, to paint, quality, rage, to sacrifice, uncle, vacation, zenith, etc.

- **Italian loanwords in English:**
  accordion, balcony, cantata, fascism, gambit, impresario, latte macchiato, mafia, novella, opera, paparazzo, quartet, ravioli, salami, tarantella, umbrella, vendetta, zucchini, etc.

- **Spanish loanwords in English**:
  albino, banana, cargo, demarcation, El Dorado, fiesta, gringo, habanera, iguana, Latino, macho, Negro, paella, Quixote, sombrero, tequila, vanilla, etc.

- **German loanwords in English**:
  ablaut, bratwurst, delicatessen, ersatz, festschrift, gestalt, kindergarten, leitmotiv, Nazi, Ossi, panzer, quartz, Realpolitik, Schadenfreude, Vaseline, Zeitgeist, etc.

- **Japanese loanwords in English**:
  aikido, bushido, geisha, haiku, judo, kamikaze, manga, Nikkei, samurai, Tamagotchi, yakuza, etc.

At present, borrowing still remains a fairly productive lexeme-building mechanism in English: of 29 new lexemes which, according to the OED, appeared in the English language between 2000 and 2011, seven were borrowed from other languages. These include:

- GALACTICO 'a skilled and celebrated footballer, especially one bought by a team for a very large fee; a football superstar' (OED / borrowed from Spanish in 2003)

- GOJI 'the edible bright red berry of either of two species of wolfberry, Lycium barbarum and L. chinense, widely cultivated in China and supposed to contain high levels of certain vitamins' (OED / borrowed from Chinese in 2002)

- HAWALADAR 'a person who acts as an agent in a hawala transaction; a hawala dealer' (OED / borrowed from Persian in 2000)

- NOROVIRUS 'a genus of caliciviruses comprising the Norwalk virus, which causes epidemic gastroenteritis in humans, and closely related caliciviruses' (OED / borrowed from scientific Latin in 2002)

- PARKOUR 'the discipline or activity of moving rapidly and freely over or around the obstacles presented by an (esp. urban) environment by running, jumping, climbing' (OED / borrowed from French in 2002)

- REGGAETON 'a form of dance music of Puerto Rican origin incorporating Latin rhythms and sounds alongside elements of dancehall reggae, hip-hop, and rap music' (OED / borrowed from Spanish in 2001)

- SUDOKU 'a type of logic puzzle, the object of which is to fill a grid of nine squares by nine squares (subdivided into nine regions of three by three squares) with the numbers one to nine, in such a way that every number appears only once in each horizontal line, vertical line, and three-by-three subdivision' (OED / borrowed from Japanese in 2000)

The OED is an excellent tool for studying not only the synchronic productivity but also the history of borrowing in English. Thus if you go to the Advanced Search page http://www.oed.com/advancedsearch, you can ask the OED to find all words that were borrowed from a particular language. Just enter e.g. 'French' to 'Language of origin' and the OED will yield all words that have ever been borrowed from French. If you enter e.g. '1900-', the OED will yield all French words which appeared in the English language between 1900 and 2011.

5.4 Lexeme-building affixation

Affixation is a major formal lexeme-building mechanism. It is different from lexeme-manufacturing and borrowing in that it involves a formal modification of the lex of an already existing lexeme. For instance, the lex of the existing lexeme TRUE was modified by means of the prefix *un-*, which gave rise to the new lexeme UNTRUE 'not faithful'.

To begin with, let us briefly recapitulate what we have already learned about affixes and affixation in the previous chapters.

- An affix is a bound morph or a bound quasi-linguistic unit.

- Affixes are shorter than roots.

- The syntactics of an affix determines its combinatory possibilities.

- Affixes can be classified into prefixes, suffixes, interfixes, and infixes.

- Affixation can be classified into isomorphic and anisomorphic affixation.

5.4.1 Affixes and their signifieds

Affixes can express a number of signifieds. For example:

- 'performer of some action': e.g. *-er* of *blogger*, *-ist* of *tourist*
- 'nationality': e.g. *-an* of *Belizean*, *-ese* of *Surinamese*

- 'gender': e.g. *-ess* of *princess*, *-ine* of *heroine*
- 'collectivity': e.g. *-dom* of *Arabdom*, *-ry* of *computery*
- 'diminutiveness': e.g. *-ie* of *mushie* 'mushroom', *-let* of *applet*
- 'location in space and time': e.g. *inter-* of *intercommunal*, *post-* of *postminimalist*
- 'pejoration': e.g. *mis-* of *miscommunicate*, *mal-* of *malware*
- 'reversal': e.g. *un-* of *unsubscribe*, *de-* of *demotivate*
- 'negation': e.g. *un-* of *untrue*, *in-* of *inadequate*

As in the case of roots, one and the same affix can be associated with more than one signified. Compare, for instance, the meanings inherent in the suffix *-er* in *blogger* and *Londoner*. In the former case, the suffix under analysis denotes a habitual performer of an action: a blogger is a person who blogs (on a more or less habitual basis). By contrast, *-er* of *Londoner* denotes an inhabitant or a native of a particular place: a Londoner is an inhabitant of London. Similarly, a New Yorker is an inhabitant or a native of New York and a Berliner is an inhabitant or a native of Berlin. Clearly, the signifieds 'performer of some action' and 'inhabitant of a particular place' are related signifieds: an inhabitant of London can be defined as a performer of the action of living in London. Nonetheless, the signifieds under analysis are not identical signifieds, so that the suffixes *-er* of *blogger* and *-er* of *Londoner* must be regarded as realizations of two different morphemes. (Analogous to uniting polysemous lexemes like TRUE$_1$ 'faithful' and TRUE$_2$ 'not false' into lexeme vocables, we can perhaps unite polysemous affixes like *-er* of *blogger* and *-er* of *Londoner* into **affix vocables**.)

5.4.2 Affixes and their syntactics

The syntactics of an affix contains two distinct kinds of information:

- **what we can do with the affix**: i.e. whether we can use the affix for new derivational formations, whether we can attach the affix to particular input signifiers, whether the affix can occupy a particular position in an output signifier.

- **what affixation will result in**: i.e. whether the addition of the affix will trigger a change of word class or whether it will impose phonetic changes on the signifier of an input lexeme.

In the following let us discuss each of these aspects. As far as the first category is concerned, we already know that the syntactics determines the combinatory properties of an affix. For example, it is the syntactics that makes the negative prefix *un-* attachable to Germanic and fully naturalized input lexes like *true*, but not to Latinate lexes like *adequate*.

A similar property concerns the ordering of an affix. As Bauer and Huddleston (2002: 1669) point out, the suffix *-hood* can be preceded but not followed by another affix: there can be a signifier like *magicianhood* but not *childhoodic. The latter is ungrammatical because the syntactics of *-hood* does not allow it to be followed by another affix.

The syntactics of an affix also determines whether it can be attached to input lexes that are members of a particular word class. As a first approximation, it may seem that this property of an affix is largely determined by its signified. Indeed, if it were not for the signified 'performer of some action', we would not be able to use the suffix *-er* for deriving deverbal nouns like *teacher*, *preacher*, *worker*, etc. denoting performers of the actions expressed by corresponding input verbs.

But consider the suffix *-eer* of nouns like *auctioneer* 'one who conducts sales by auction' (OED / 1708) and *sonneteer* 'a composer of sonnets' (OED / 1667). According to the OED, the suffix *-eer* carries the meaning 'one who is concerned with, one who deals in'. This suffix is still productive in Present-day English. According to Word Spy, the most recent neologisms involving this suffix are:

- PUMPKINEER 'a person who grows giant pumpkins, particularly ones meant to be entered in pumpkin-weighing contests' (Word Spy / 1988)

- FUSIONEER 'a person who investigates nuclear fusion or attempts to build a nuclear fusion reactor' (Word Spy / 1988).

According to the OED, the most recent *-eer* formations are:

- FREE MARKETEER 'a proponent of, or believer in, the free market' (OED / 1963)

- SUMMITEER 'one who takes part in summit meetings' (OED / 1957)

- WEAPONEER 'one who has charge of a weapon of war prior to its deployment' (OED / 1945)

- ROBOTEER 'a person who designs or constructs robots; an expert in robotics' (OED / 1930)

- RACKETEER 'a person (especially a member of a gang or crime syndicate) who earns money through a dishonest or illegal business, typically involving extortion, intimidation, or violence' (OED / 1924)

- FICTIONEER 'a writer or inventor of fiction' (OED / 1923)

- SLOGANEER 'one who devises or who uses slogans' (OED / 1922)

Given these formations, we can conclude that *-eer* can only attach to nominal input lexes. That is:

- *pumpkineer* ← *pumpkin* + *-eer*
- *fusioneer* ← *fusion* + *-eer*
- *free marketeer* ← *free market* + *-eer*
- *summiteer* ← *summit* + *-eer*
- *weaponeer* ← *weapon* + *-eer*
- *roboteer* ← *robot* + *-eer*
- *racketeer* ← *racket* + *-eer*
- *fictioneer* ← *fiction* + *-eer*
- *sloganeer* ← *slogan* + *-eer*

Evidently, this fact cannot only be attributed to the signified of the suffix *-eer* 'one who is concerned with'. We can easily imagine a person who is concerned with teaching (without, at the same time, being a teacher) or a person who is concerned with blogging (without, at the same time, being a blogger). However, it is extremely unlikely that an English speaker will ever call such people **teacheer* and **bloggeer*. The reason for this is that the syntactics of *-eer* does not allow attaching it to non-nominal input lexes.

The last aspect pertaining to combinatorial properties of an affix is its productivity, i.e. the ability to attach to new input lexes (i.e. to those lexes it has not been previously attached to), thus creating new lexemes. One **conditio sine qua non** (i.e. a necessary condition) of the productivity of an affix is that of being a **living affix** (Gizburg et at. 1979: 123), i.e. being recognizable as a component part of complex lexes in which it occurs. As an illustration of this point, let us consider the signifiers *dead* /dɛd/ and *death* /dɛθ/. Since these semantically related signifiers differ only with regard to the final consonant sound, it can be conjectured that /d/ of *dead* is an adjective-forming suffix carrying the meaning 'quality' – *dead* means 'the quality of being dead' – and /θ/ of *death* is a noun-building suffix carrying the nominal meaning 'state': death is the state of being *dead*.[14] This analysis is true from a diachronic point of view. According to the OED, /d/ of *dead* is indeed a suffix that goes back to Germanic *daudo-z*, whereas /θ/ of *death* goes back to Germanic *dauþu-z*. However, today there do not seem to be speakers of English who do indeed segment the word *dead* into the root /dɛ/, carrying the meaning 'dead', and the suffix /d/, carrying

---

[14] The analysis of /θ/ of *death* as a suffix meaning 'state' seems to be supported by words like *truth*, where /θ/ is indeed a suffix carrying the meaning 'state': truth is the state of being true.

the meaning 'quality'. Similarly, the word *death* does not seem to be synchronically segmentable into the root /dɛ/ and the suffix /θ/, carrying the meaning 'state'. Both *dead* and *death* are monomorphemic words whose morphs *dead* and *death* carry the meanings 'dead' and 'death'. /d/ of *dead* and /θ/ of *death* are thus **dead affixes** and, accordingly, cannot be used for new derivational formations. That is, there cannot be new formations like **gooth* meaning 'the state of being good' (from the adjectival input lex *good*, where /d/ is an affix carrying the adjectival meaning 'quality') and **reth* meaning 'the state of being red' (from the adjectival input lex *red*, where /d/ is an affix carrying the adjectival meaning 'quality'). Since /d/ of *dead* is a dead affix, which is not recognized as a component part of the signifier *dead*, it cannot give rise to the re-analysis of the originally monomorphemic words *good* and *red* as combinations of the adjectival roots /gʊ/ and /rɛ/, carrying the meanings 'good' and 'red', and the suffix /d/, carrying the meaning 'quality'.

However, note that not all living affixes are productive affixes. Consider, for instance, the affix /θ/ of forms like *strength*, *length*, *depth*, etc. Despite the obvious idiomaticity of the signifieds of these lexemes – e.g. *strength* does not really mean 'the state of being strong' but 'a scale with regard to which a person can be described as strong or weak' – /θ/ of *strength*, *length*, *depth*, etc. is undoubtedly a recognizable component of these signifiers (and can thus be regarded as at least a morfoid). However, despite its living character, the affix /θ/ is currently not used for new derivational formations. According to the OED, the last formations involving this suffix are *illth* 'ill-being' (1862) and *sidth* 'length; depth' (1855). The living suffix /θ/ is currently an unproductive derivational suffix, whose syntactics does not allow it to be used for new derivational formations. Hence we can argue for the impossibility or at least the extreme unlikelihood of formations like **goodth, redth*, badth**.

A similar example of a synchronically living but unproductive affix is the suffix -*ant* of words like *defendant* and *accountant*. According to the OED, the last formations involving this suffix took places in the 1970s and the 1960s. These include:

- REASSORTANT 'of a virus: having a genome consisting of parts derived from the genomes of two (or more) different viruses, as a result of reassortment during mixed infection of the same host or cells' (OED / 1979)

- TRANSCONJUGANT 'a plasmid or a bacterial cell which has received genetic material by conjugation with another bacterium' (OED / 1974)

- RECORDANT 'a person who records a trademark with the customs authorities in order to help prevent the importation of goods that infringe the mark' (OED / 1969)

- PROPPANT 'a granular material which is pumped with a fluid medium under pressure into rock, so that fractures formed in this process are held open when the pressure is released, allowing oil or gas to flow more freely' (OED / 1966)

- TRANSDUCTANT 'a cell into which genetic material has been transduced' (OED / 1963)

- DOPANT 'the substance used in doping a semiconductor' (OED / 1963)

- INCAPACITANT 'a substance that can be used to incapacitate a person for a time without wounding or killing him' (OED / 1961)

- ANOVULANT 'a drug or other agent that suppresses ovulation' (OED / 1960)

In addition to these lexemes, the OED also regards the 1983 formation EXFOLIANT 'a cosmetic product designed to remove dead cells from the surface of the skin' as a product of affixation of the verb *exfoliate* by means of the suffix *-ant*. At the same time, the OED acknowledges that this form could have been borrowed from French: cf. the English *exfoliant* and the French *exfoliante*. Anyway, since the early 1980s the suffix *-ant* has not been used for new derivational formations and can thus be regarded as a non-productive affix.

Now, let us proceed to consequences of affixation, i.e. the questions of whether the addition of a particular affix will result in a change of word class and whether it will trigger phonetic changes of the lex of an input lexeme.

What is meant by the former question is that while e.g. the addition of the suffix *-er* to verbal input lexes like *read* produces nominal output lexemes (READER, BLOGGER), the addition of the negative prefix *un-* to adjectival input lexes like *true* produces adjectival output lexemes (UNTRUE, UNSACKABLE). A change of word class is usually brought about by affixation (cf. TEACH and TEACHER, CLOUD and CLOUDY, BELIZE and BELIZEAN, etc.), whereas prefixation usually produces output lexemes whose lexes are members of the same word class as lexes realizing their input lexemes: cf. TRUE and UNTRUE, COMMUNAL and INTERCOMMUNAL, COMMUNICATE and MISCOMMUNICATE. At the same time, note that there are word class non-changing suffixes like *-eer* of *auctioneer*: both the input lexeme AUCTION and the output lexeme AUCTIONEER are nouns. And there are word class changing prefixes: for example, while the input lex *friend* is a noun, the output prefixed lex *to befriend* 'to act as a friend to, to help, favor; to assist, promote, further' (OED) is a verb.

At first glance, it may be tempting to attribute this important aspect of the syntactics of an affix to its signified. Indeed, if *-er* means 'performer of some action', then the addition of this suffix to verbal input lexes like *to read* must give rise to nominal output lexes like *reader*. Similarly, if *-y* means 'full', then the addition of this suffix to a nominal input lex like *cloud* must give rise to an adjectival output lex like *cloudy*. However, the main reason why the lex *reader* is a noun and the lex *cloudy* is an adjective is that both these output lexes (or to be more precise, their signifieds) are headed by the suffixes *-er* and *-y*. That is, the output signified 'teacher' can be represented as the NP *performer of the action of teaching*, whose head is the nominal meaning component 'performer', inherent in the affix *-er*. Similarly, the output signified 'cloudy' can be represented as the adjective phrase *full of clouds*, whose head is the adjectival meaning component 'full', inherent in the suffix *-y*. The nominal meaning 'performer' and the adjectival meaning 'full' are thus the **head meanings** of the output lexemes READER and CLOUDY.

By contrast, the head meaning of the output lexeme MISCOMMUNICATE is not the meaning 'pejoration', inherent in *mis-*, but the meaning 'to communicate', inherent in *communicate*: *miscommunicate* means 'to communicate badly'. This is the reason why the addition of the prefix *mis-* to the input lex *communicate* gives rise to the output lex *miscommunicate*, which is a member of the same word class as the input lex *communicate*.

The signified 'auctioneer' is similar to the signifieds 'teacher' and 'cloudy' in that it is also headed by the suffix *-eer*. That is, the head meaning of the signified 'auctioneer' is the nominal meaning 'person', inherent in the affix *-eer*. However, since in *auctioneer* this meaning combines with the nominal meaning 'auction' – *auctioneer* means 'a person who conducts auctions' – the suffix *-eer* seems similar to the prefix *mis-* in that it has also produced the output lexeme which is a member of the same word class as the corresponding input lexeme. The suffix *-eer* of *auctioneer* is, however, different from the prefix *mis-* of *miscommunicate* in that the former heads the output lexeme AUCTIONEER.

In summary, the syntactics of an affix also contains the information as to whether the affix can function as head of a new lexeme. This information determines whether the meaning inherent in the affix will be able to function as head meaning of the signified of an output lexeme, thereby (often, but not always) giving rise to an output lex which is a member of a different word class than a corresponding input lex.

As regards the ability to trigger phonetic changes, let us compare the following three formations:

- CLOUDY /ˈklaʊdɪ/; cf. the input lex /klaʊd/
- BAZILLIONAIRE /bəˌzɪljəˈnɛː/; cf. the input lex /bəˈzɪljən/
- ADDITIONALITY /əˌdɪʃəˈnalɪti/; cf. the input lex /əˈdɪʃn(ə)l/

These examples serve to illustrate that affixes can be classified into the following three categories:

1. **stress-neutral affixes**
2. **stress-attracting affixes**
3. **stress-shifting affixes**

The adjective-building suffix *-y* of *cloudy* is an example of a stress-neutral affix. As one can notice, the output signifier *cloudy* is stressed on the first syllable, just as the input signifier *cloud*.

The noun-building suffix *-aire* of *bazillionaire* 'a person of enormous wealth' (OED / 1987) is an example of a stress-attracting affix: the primary stress of *bazillionaire* falls on the suffix *-aire*, not on the second syllable as in the input signifier /bəˈzɪljən/.

Finally, the suffix *-ity* of *additionality* is an example of a stress-shifting suffix: its addition to the input signifier *additional* has shifted the primary stress from the second to the fourth syllable: cf. /əˌdɪʃəˈnalɪti/ and /əˈdɪʃnˌ(ə)l/.

Stress-neutral affixes also include the majority of prefixes as well as the suffixes *-al -er*, *-ist*, *-ship*, etc. Compare, for example:

- /sɪˈmɛtrɪk/ and /æsɪˈmɛtrɪk/ (*symmetric* and *asymmetric*)
- /kəˈmjuːnɪkeɪt/ and /ˌmɪskəˈmjuːnɪkeɪt/ (*communicate* and *miscommunicate*)
- /ˈsɛntrəlaɪz/ and /diːˈsɛntrəlaɪz/ (*centralize* and *decentralize*)
- /əˈkjuːz/ and /əˈkjuːzəl/ (*accuse* and *accusal*)
- /tiːtʃ/ and /ˈtiːtʃə(r)/ (*teach* and *teacher*)
- /dɪˈnæmɪks/ and /daɪˈnæmɪsɪst/ (*dynamics* and *dynamicist*)
- /ˈdəʊnə(r)/ and /ˈdəʊnəʃɪp/ (*donor* and *donorship*)

Stress-attracting affixes also include the suffixes *-ation*, *-eer*, *-ese*, *-esque*, *-ette*, etc. For example:

- /ˈɛskəleɪt/ and /ˌɛskəˈleɪʃən/ (*escalate* and *escalation*)
- /ˈɔːkʃən/ and /ɔːkʃəˈnɪə(r)/ (*auction* and *auctioneer*)
- /s(j)ʊərɪˈnæm/ and /s(j)ʊərɪnæˈmiːz/ (*Suriname* and *Surinamese*)
- /kəmˈpjuːtə/ and /ˌkəmpjuːtəˈrɛsk/ (*computer* and *computeresque*)
- /ˈpɛdəgɒg/ and /ˌpɛdəgʊˈgɛt/ (*pedagogue* and *pedagoguette*)

Finally, stress-shifting affixes also include *-ic*, *-ial*, *-ual*, etc. For example:

- /ˈsɛljʉlʌɪt/ and /ˌsɛljʉˈlɪtɪk/ (*cellulite* and *cellulitic*)
- /kəmˈpəʊnənt/ and /kɒmpəˈnɛnʃɪəl/ (*component* and *componential*)
- /ˈkɒnflɪkt/ and /kənˈflɪktjuːəl/ (*conflict* and *conflictual*)

Like synchronic productivity, the status of an affix as a stress-neutral, stress-attracting, or stress-shifting affix is not predictable from either its signifier or its signified. Also, there does not seem to exist a purely phonetic explanation (such as e.g. a place-of-articulation assimilation) for the fact that, for example, the addition of the suffix *-aire* to *bazillion* must necessarily shift the stress from the second syllable /zɪ/ to the fourth syllable /nɛː/.[15] Nouns in English are usually stressed on the **penultimate syllable** (i.e. the pre-final syllable) provided that it is **heavy**, i.e. a syllable which ends in a consonant, a diphthong, or a long vowel. When the penultimate syllable is **light** (i.e. a syllable ending in a short vowel), the primary stress moves to the **antepenult** (i.e. the syllable preceding the **penult**). For example, *algebra* /ˈældʒɪbrə/ is stressed on the antepenult /æl/ because the penult /dʒɪ/ is a light syllable: it ends in the short vowel /ɪ/. Now, let us consider the placement of the primary stress in /bəˌzɪljəˈnɛː/. Given that the penultimate syllable /jə/ is light, one would expect *bazillionaire* to retain the primary stress on the second syllable /zɪl/. This, however, does not happen because, as said above, the suffix *-aire* is a stress-attracting suffix, i.e. its syntactics orders English speakers to place the primary stress on *-aire*, whenever this suffix is used for derivational formations.

5.4.3 Productivity

Affixation is a very productive lexeme-building mechanism in Present-day English. Thus, of 587 new signifiers which, according to the OED, appeared in the English language between 1990 and 2011, 170 can be regarded as instances of affixation. Below are just some of the examples:

- ANORAKY (← *anorak* + *-y*) 'boring, overly studious, unfashionable, or socially inept; specially displaying obsessive or fastidious concern with the details of a hobby or special interest […]' (OED / 1992)

- BLAIRISM (← *Blair* + *-ism*) 'the policies and principles advocated by Tony Blair' (OED / 1994)

- CABBAGED (← *cabbage* + *-ed*) 'incapacitated by drugs or alcohol (or their after-effects); extremely intoxicated' (OED / 1991)

---

[15] Note that this does not apply to all phonetic changes triggered by the addition of a derivational affix. For example, as Halle (2005) argues, so-called **velar softening** exemplified by the change of the velar consonant [k] of e.g. /aɪˈkɒnɪk/ into the alveolar [s] of e.g. /aɪkəˈnɪsɪtɪ/ is an instance of place-of-articulation assimilation of the velar [k] to the front vowel [ɪ] rather than an inherent property of the syntactics of the noun-building affix *-ity*.

- DOWNSHIFTER (← *to downshift* + *-er*) 'a person who adopts a less pressured and demanding career or lifestyle, especially one who accepts a reduced income in pursuit of personal fulfillment' (OED / 1990)

- EEYORISH (← *Eeyore* + *-ish*) 'deeply pessimistic; gloomy' (OED / 1992)

- GENOMICIST (←*genomic* + *-ist*) 'a scientist who works in the field of genomics' (OED / 1995)

- LADETTE (← *lad* + *-ette*) 'a young woman characterized by her enjoyment of social drinking, sport, or other activities typically considered to be male-oriented [...]' (OED / 1995)

- TO OVERSUPINATE (← *over-* + *to supinate*) 'to run or walk so that the weight falls upon the outer sides of the feet to a greater extent than is necessary, desirable, etc.; to supinate excessively' (OED / 1990)

- TO REWILD (←*re-*+ *to wild*) 'to return (land) to a wilder and more natural state' (OED / 1990)

To study the productivity of a particular suffix (e.g. *-ish* of *Eeyorish*), go to the OED Advanced Search page http://www.oed.com/advancedsearch and enter '*ish' (without quotation marks) to 'Restrict to entry letter or range'. The OED will then yield all lexemes whose lexes end in *-ish*. Some of them will be adjectives like *Eeyorish* which contain the suffix *-ish*. You can refine your search by entering e.g. '1990-' to 'Date of entry' and choosing the option 'Adjective' in 'Part of speech'. The OED will then yield all recent adjectival lexemes (i.e. which appeared between 1990 and 2011) whose lexes end in *-ish*.

To study the productivity of a particular prefix (e.g. *re-* of *rewild*), go to the above mentioned OED Advanced Search page and enter 're*' (without quotation marks) to 'Restrict to entry letter or range'. The OED will then yield all lexemes whose lexes begin with *re-*. Some of them will be verbs like *to rewild* which contain the prefix *re-*. You can refine your search by entering e.g. '1990-' to 'Date of entry' and choosing the option 'Verb' in 'Part of speech'. The OED will then yield all recent verbal lexemes (i.e. which appeared between 1990 and 2011) whose lexes begin with *re-*.

As was pointed out in 4.2.8, the synchronic productivity of a derivational affix can also be studied with the help of Word Spy. Just go to the Word Spy Advanced Search page http://www.wordspy.com/search.asp, enter '*ish' or 're*' (both without quotation marks) to the search mask, choose the option 'Within title', and then click at 'Search'. Word Spy will yield all words ending in *-ish* and beginning with *re-*. Some of them could be adjectives like *Eeyorish* and verbs

Lexeme-building mechanisms 143

like *to rewild*. Unfortunately, it is not possible to instruct Word Spy to search for words that are members of a particular word class.

5.5 Lexeme-building apophony

Lexeme-building apophony is any modification of the lex of an input lexeme that does not qualify as an instance of segmental affixation. One type of apophony (which has been mentioned in 4.2.2) is the **stress shift**. Consider, for example, the following formations:

- *to increase* /ɪnˈkriːs/ → *an increase* /ˈɪnkriːs/
- *to insult* /ɪnˈsʌlt/ → *an insult* /ˈɪnsʌlt/
- *to permit* /pəˈmɪt/ → *a permit* /ˈpəːmɪt/

In all these cases, the stress shift is accompanied by quasi-idiomatization of the input signifieds, i.e. an increase is the process of increasing; an insult is the act of insulting; a permit is a document that permits something (e.g. a residence permit is a document that grants permission to reside in some country).

In addition to the stress shift, apophonies also include:

- **consonant change**
- **vowel change**
- **consonant change accompanied by vowel change**
- **stress shift accompanied by vowel change**

As an illustration of the first category, let us again consider the pair *dead–death*, which we discussed in the previous section. As we established, neither [d] of *dead* nor [θ] of *death* can be regarded as an affix from a synchronic point of view. Given this fact and given that the signified 'death' can be analyzed as a quasi-idiom in relation to the signified 'dead' – death is the state or condition of being of dead – we are justified in concluding that the lexeme DEATH is a product of apophony of the lexeme DEAD: the former came into existence via the change of the final consonant [d] of the input lex *dead* to [θ] in the output lex *death*. Similar examples include:

- *to defend* /dɪˈfɛnd/ → *defense* /dɪˈfɛns/
- *to speak* /spiːk/ → *speech* /spiːtʃ/
- *to believe* /bɪˈliːv/ → *belief* /bɪˈliːf/

As an illustration of the second category, let us consider the pair *full* /fʊl/–*to fill* /fɪl/. From a semantic point of view, the signified of the verb *to fill* can be analyzed as a quasi-idiom in relation to the signified of the adjective *full*: *to fill* can be said to mean 'to make full'. From a formal point of view, the lexes of the two lexemes differ only with regard to the vowel sounds [ʊ] of *full* and [ɪ] of *fill*. Neither the former nor the latter can be regarded as an affix carrying a discernible meaning of its own: neither *full* nor *fill* seem to be synchronically segmentable into the root /fl/, carrying the meaning 'full', and the infixes [ʊ] / [ɪ], carrying the meaning 'quality of being [...]' / 'process of making [...]'. Both lexes are monomorphemic words, which cannot be segmented into smaller units (be it morphs or quasi-linguistic units). Accordingly, we can conclude that the lexeme TO FILL is a product of apophony of the lexeme FULL: the former came into existence via the change of the root vowel [ʊ] of the input lex *full* to [ɪ] in the output lex *fill*. Similar examples include:

- *blood* /blʌd/ → *to bleed* /bli:d/
- *separate* /ˈsɛpərət/ → *to separate* /ˈsɛpəreɪt/
- *legitimate* /lɪˈdʒɪtɪmət/ → *to legitimate* /lɪˈdʒɪtɪmeɪt/

The third category is essentially the blend of the first two categories. Consider, for instance, the pair *life* /lʌɪf/–*to live* /lɪv/. As in the previous examples, there exists a quasi-idiomatic relation holding between the signifieds of the lexemes in question: *to live* can be said to mean 'to be alive or to possess life'. As regards the formal side, the difference concerns the last two sounds /ʌɪf/ of *life* and /ɪv/ of *live*. Similar to previous examples, we cannot regard these sounds as affixes. That is, the noun *life* cannot be segmented into the root /l/ and the suffix /ʌɪf/ and the verb *live* cannot be segmented into the root /l/ and the suffix /ɪv/. (Recall that roots are usually longer than affixes.) Given this fact, we are justified in concluding that the lexeme TO LIVE is a product of apophony of the lexeme LIFE: the former came into existence via the change of the root vowel [ʌɪ] to [ɪ] as well as via the change of the final consonant [f] to [v].

Finally, the last category is the blend of the categories 'stress shift' and 'vowel change'. Examples include such pairs as:

- *conduct* /ˈkɒndəkt/ → *to conduct* /kənˈdʌkt/
- *contest* /ˈkɒntɛst/ → *to contest* /kənˈtɛst/
- *fragment* /ˈfrægmənt/ → *to fragment* /frægˈmɛnt/

Here the change of the stress is accompanied by the change of the quality of one or more vowels. Strictly speaking, the latter represents a consequence of the former: it is a well-known fact that vowels in unstressed syllables often undergo reduction and change to a **schwa** (i.e. the sound /ə/). That is why this last type of apophony represented by pairs like *conduct–to conduct* can also be regarded as an instance of the category 'stress shift' rather than as an independent category.

5.5.1 Productivity

In Present-day English apophony is a much less productive lexeme-building mechanism than affixation, borrowing, and semantic change: of 587 new signifiers which, according to the OED, appeared in the English language between 1990 and 2011, no more than three can be regarded as instances of lexeme-building apophony. These are the lexes that were already named in 4.2.8:

- *adultescent* /ˌadʌlˈtɛsnt/ ← *adolescent* /ˈædəʊˈlɛsənt/
- *Google* /ˈguːgl/ ← *googol* /ˈguːgɒl/
- *Lollywood* /ˈlɒlɪwʊd/ ← *Bollywood* /ˈbɒlɪwʊd/

Earlier examples of lexeme-building apophony include:

- *def* (← *death*) 'excellent, outstanding; fashionable, cool' (OED / 1981)

- *herstory* (← *history*) '[...] history emphasizing the role of women or told from a woman's point of view; also, a piece of historical writing by or about women' (OED / 1970)

- *to wazz* (←*to whizz* 'to make a sound as of a body rushing through the air') 'to urinate' (OED / 1984)

To study the productivity of apophony, go to the OED Advanced Search page http://www.oed.com/advancedsearch, enter 'alteration' (without quotation marks) to the first row of the search mask from above, and choose the option 'Etymology' instead of 'Full-text' (near the search mask). The OED will then yield all entries whose Etymology-sections contain the word *alteration* (i.e. how the OED refers to phonetic changes which this book regards as instances of apophony). Some of these entries will be devoted to lexemes whose lexes came into existence via apophony.

5.6 Compounding

Compounding is the creation of a new lexeme by means of combining at least two input roots into the lex of a new complex lexeme, a **compound lexeme**. For example, PLAYGROUND ← *play* + *ground*.

Note that in English a considerable number of compound lexemes, whose lexes qualify as compounds from a synchronic point of view (i.e. they are synchronically segmentable into at least two roots), do not qualify as instances of compounding from a diachronic perspective. For example, as we have learned in 4.2.4, the compound nominal lexeme BOYFRIEND$_2$ 'a male sexual partner', whose lex is synchronically segmentable into the roots *boy* and *friend*, came into existence via metonymy-based full-idiomatization of the lexeme BOYFRIEND$_1$ 'a boyhood friend'. Similarly, the compound verbal lexeme BABYSIT came into existence via back-formation of the nominal lexeme BABYSITTER.

Some authors regard compound lexemes like BOYFRIEND$_2$ and BABYSIT as **pseudo-compounds**. The distinction between a pseudo-compound like BABYSIT, whose lex is a product of another lexeme-building mechanism, and a **genuine compound** like PLAYGROUND, whose lex is indeed a product of compounding, is of particular theoretical importance for those students of English morphology who investigate the mechanism of compounding (e.g. differences between compounding and other lexeme-building mechanisms). Indeed, if you are interested in the peculiarities of this process, you must ignore pseudo-compounds like BOYFRIEND$_2$ and BABYSIT: these do not qualify as instances of compounding and, accordingly, must be left out of consideration.

Since the goal of this section is to describe the mechanism of compounding, we will accept the distinction between a pseudo-compound and a genuine compound. To find out whether the compound under analysis is indeed a product of compounding, we will rely on the OED. For example, the Etymology-section of the OED entry for PLAYGROUND contains the following information: 'Etymology < *play* n. + *ground* n'. Relying on this information, we can conclude that *playground* is indeed an instance of compounding, i.e. a compound lex which was created via combining the two input roots *play* and *ground* into the new compound lex *playground*.

5.6.1 Compounding as an anisomorphic mechanism

As we argued in 4.2.3, compounding is an anisomorphic lexeme-building mechanism, i.e. a mechanism that produces output lexemes whose signifieds are not (or not entirely) representable in terms of their components' signifieds. Thus of 81 recent instances of pure compounding (of 587 new signifiers which appeared in the English language between 1990 and 2011), there are no isomorphic compounds, i.e. compound lexemes whose signifieds are made up of their components' signifieds only. The overwhelming majority of the recently formed compounds – 50 compounds – exhibit quasi-idiomatic signifieds. For example:

- DRUM AND BASS 'a style of popular dance music originating in Britain in the early 1990s, variously thought of as derived from or identical to jungle, and characterized primarily by a fast drum track and a heavy, usually slower, bass track, but often also featuring synthesized or sampled strings, piano, or other instrumentation' (OED / 1992)

- STAR 69 'a call return service which automatically dials the number from which the last incoming call was made, activated by dialing the 'star' (*), 6, and 9 buttons on a touch-tone phone' (OED / 1990)

- INFORMATION FATIGUE 'apathy, indifference, or mental exhaustion arising from exposure to too much information, especially (in later use) stress induced by the attempt to assimilate excessive amounts of information from the media, the Internet, or at work' (OED / 1991)

- NO-MATES 'a person (usually a man) regarded as lonely or having no friends' (OED / 1993)

- YEAR 2000 'designating or relating to computer problems arising from the inability of certain software and firmware to deal correctly with dates of 1 January 2000 or later, owing to the numerical representation of calendar years by the last two digits only' (OED / 1993)

- TO EGO-SURF 'to search on the Internet for mentions of one's own name or the name of one's business, website, etc.' (OED / 1995)

- EURO NOTE 'any of various banknotes denominated in the euro currency' (OED / 1995)

- DOTCOM 'an Internet address for a commercial site expressed in terms of the formulaic suffix *.com*; a web site with such an address' (OED / 1994)

- ALWAYS-ON (adjective) 'relating to or designating a continuously accessible connection to the Internet, especially in contrast to one that requires activation by dial-up or other means' (OED / 1996)

21 compounds have semi-idiomatic signifieds. For example:

- HOT DESK 'a shared office desk or workstation, occupied on a temporary, ad hoc basis or part-time basis, and not allocated permanently to an individual' (OED / 1990)

- GENDERQUAKE 'a radical alteration in the relationship between the sexes, especially one resulting from deliberate changes in women's economic or political activity' (OED / 1993)

- DADROCK 'rock music that appeals to an older generation, or is heavily influenced by that of an earlier era, especially the 1960s' (OED / 1994)

- MEATSPACE 'the physical world, as opposed to cyberspace or a virtual environment' (OED / 1995)

- BLUE STATE 'a state (projected to be) won by the Democratic candidate in a presidential election. More generally: a Democratic state; a state which tends to vote Democrat' (OED / 2000)

Finally, only 10 compounds can be analyzed as full-idioms. For example:

- TO DRAG AND DROP 'to move or copy (an image, icon, text, etc.) from one part of a display screen to another using a mouse or similar device using a drag-and-drop facility' (OED / 1990)

- CARPET MUNCHER 'a lesbian' (OED / 1992)

- DOWN-LOW (adjective) 'secret, quiet, low-profile; (in later use) specially of or relating to men who secretly engage in homosexual activity' (OED / 1991)

These numbers tell us that quasi-idiomatization is the default semantic outcome of compounding. That is, the majority of lexemes that come into existence via compounding have signifieds that contain their components' signifieds plus some additional idiomatic signifieds. The reason for this is that the creation of a quasi-idiomatic compound requires a lesser effort on the part of a language user than does the creation of a semi- or fully-idiomatic compound. The latter – especially metaphor-based full-idioms like CARPET-MUNCHER and DOWN-LOW – require a considerable creativity and therefore do not occur as often as quasi-idiomatic compounds.

5.6.2 The semantics of compounding

Since compounding is an anisomorphic lexeme-building mechanism, which always produces idiomatic lexemes, any study of compounding must try to answer the question of whether there are any regularities governing the creation of quasi-, semi-, and fully-idiomatic compound signifieds.

To begin with, it must be observed that everything we have learned about semantic change in Section 5.1 is true of anisomorphic compounding. This means that the compounding of the lexes of two or more input lexemes into the lex of a new compound lexeme can be accompanied by either metonymization or metaphorization of one or both of its components' signifieds. For instance, the semi-idiom DADROCK 'rock music that appeals to an older generation' is a product of part-for-whole metonymization of the signified 'dad': our fathers (alongside with mothers, grandmothers, grandfathers, etc.) are representatives of an older generation. By contrast, the semi-idiom GENDERQUAKE 'a radical alteration in the relationship between the sexes' came into existence via metaphorization of the signified 'quake': a radical alteration in the relationship between the sexes was metaphorically analogized to a physical quake.

Similarly, fully-idiomatic compounds may come into existence via:

1. **metaphorization of all of their components' signifieds**
2. **metonymization of all of their components' signifieds**
3. **combination of these two strategies**

For example, the compound *carpet muncher* 'a lesbian' came into existence via metaphorization of the input signifieds 'carpet' and 'muncher': lesbian sex (especially cunnilingus) must have been metaphorized as munching a carpet. This analysis is corroborated by an earlier formation RUG MUNCHER 'a lesbian. Also (occasionally): a man who performs cunnilingus' (OED / 1981): in this formation female genitalia are metaphorized as a rug on which a man performing cunnilingus munches.

As an illustration of a metonymy-based fully-idiomatic compound, consider the lexeme GREEN ACCOUNTING 'a system in which economic measurements take into account the effects of production and consumption on the environment' (Word Spy / 1989). This compound came into existence via metonymization of the input signifieds 'green' and 'accounting': the green color is a well-known symbol of environmental issues, so that the input signified 'green' can be said to metonymically stand for the output signified 'environment'; similarly, economic measuring can be seen as an instance of accounting, so that the input signified 'accounting' can be said to metonymically stand for the output signified 'economic measurements'.

Finally, consider the fully-idiomatic lexeme GREY NOMAD 'a retired person who travels extensively, particular in a recreational vehicle' (Word Spy). This compound came into existence via metonymization of the input signified 'grey' and metaphorization of the input signified 'nomad'. As regards the former, old people often have grey hair, so that the input signified 'grey' can be said to metonymically stand for the idiomatic output signified 'old or retired people'. As for the latter, the input signified 'nomad' cannot be regarded as a metonym for the output signified 'one who travels extensively, particular in a recreational vehicle': a person who extensively travels in a recreational vehicle is not literally a nomad, i.e. 'a member of a people that travels from place to place to find fresh pasture for its animals, and has no permanent home' (OED). The former can only be metaphorically analogized to the latter: that is, the activity of extensively travelling in a recreational vehicle can be perceived of as being similar to the travelling activities of nomads.

As regards quasi-idiomatic compounds, it appears that they fall into two main categories:

1. **quasi-idiomatic compounds of the *information fatigue*-type**
2. **quasi-idiomatic compounds of the *drum and bass*-type**

Quasi-idiomatic compounds of the *information fatigue*-type are quasi-idioms like INFORMATION FATIGUE whose additional idiomatic meanings make these compounds' signifieds narrower than the mere sum of their components' signifieds. That is, information fatigue is a particular kind of fatigue that has something to do with information: fatigue that arises from exposure to too much information; a euro note is a particular banknote that has something to do with the euro currency: a banknote which is denominated in the euro currency; ego-surfing is a particular instance of surfing the Internet that has something to do with oneself: surfing the Internet for mentions of one's own name; etc. With regard to the quasi-idiomatic compounds that were discussed in the previous chapters, this pattern is characteristic of e.g. the compounds *spring festival* and *brake cable*: as we argued, a spring festival is a particular festival that has something to do with spring and a brake cable is a particular cable that has something to do with a brake.

Quasi-idiomatic compounds of the *drum and bass*-type are quasi-idioms like DRUM AND BASS whose components name some (more or less) important characteristics of what the lexes realizing these lexemes are used to refer to. Thus drum and bass is a style of dance music which is characterized by a fast drum track and a slower bass track. Similarly, star 69 is a service that is activated by dialing '*69'; a No-Mates is a lonely man who has no mates; a dotcom is an Internet address that is expressed in terms of the formulaic suffix '.com'; etc. With regard to the previously discussed quasi-idiomatic compounds, this pattern can be found in FOOTBALL, BASKETBALL, HANDBALL, etc. As we established in 2.4.7, the lexes of these lexemes are segmentable into the components *foot / ball*, *basket / ball*, *hand / ball*, *volley / ball*, etc., whose signifieds provide an (incomplete) explanation for how these games are supposed to be played: a football is a game that involves a ball and the players' feet; basketball is a sport that involves a ball and a basket; handball is a sport that involves a ball and the players' hands; etc.

Quasi-idiomatic compounds of the *drum and bass*-type are sometimes called **bahuvrihi compounds**, from the Sanskrit *bahuvrihi* 'having much rice' (OED). The term 'bahuvrihi' has usually been used in connection with compounds like SKINHEAD 'a person whose hair is worn very short or shaved off entirely' (OED) and CUTTHROAT 'one who cuts throats; a ruffian who murders or does deeds of violence [...]' (OED) which denote individuals possessing the characteristics named by these compounds' components. That is, a skinhead is a person whose hair is worn very short, so that other people can see the skin of his head. And a cutthroat is a murderer who murders by cutting other people's throats. In other words, the components *skin* and *head* name an important characteristic of a skinhead and the components *cut* and *throat* name an important characteristic of a cutthroat. It is not difficult to notice that the bahuvrihis SKINHEAD and CUTTHROAT exhibit the very same semantic pattern as the quasi-idiomatic compounds DRUM AND BASS, STAR 69, NO-MATES, DOTCOM, FOOTBALL, etc.

In addition to classifying quasi-idiomatic compounds into these two rather general categories, one may also try to discover the most frequently recurrent patterns of quasi-idiomatization in compounding. This issue has been recently addressed by Jackendoff (2009: 123-124), who, without using the term 'quasi-idiom', has proposed "a list of the (most prominent) **basic functions** for English compounds". What Jackendoff calls 'a basic function' is, however, nothing more than just an idiomatic meaning that is a part of a quasi-idiomatic compound's signified which cannot be attributed to either of its components' signifieds. These include, for example:

- the meaning 'both' in compounds like POLITICIAN-TYCOON: a politician-tycoon is a person who has the properties of both a politician and a tycoon.

- the meaning 'serves as' in compounds like BUFFER STATE: a buffer state is a state that serves as a buffer.

- the meaning 'located at / in / on / near' in compounds like WINDOW SEAT: a window seat is a seat that is located near a window.

- the meaning 'takes place at specified time' in compounds like SPRING FESTIVAL: a spring festival is a festival that takes place in spring.

- the meaning 'caused by' in compounds like KNIFE WOUND: a knife wound is a wound that was caused by a knife.

- the meaning 'consists of' in compounds like BRASS INSTRUMENT: a brass instrument is an instrument that consists of brass.

- the meaning 'part of' in compounds like BRAKE CABLE: a brake cable is a cable that is a part of a brake.

- the meaning 'made by' in compounds like FOOTPRINT: a footprint is a print made by a foot.

- the meanings 'protect something' and 'protect from something' in compounds like CHASTITY BELT and SUN HAT: a chastity belt is a belt that protects chastity and a sun hat is a hat that protects from the sun.

Despite the fact that many quasi-idiomatic compounds do indeed come to signify these 'basic' meanings, the semantic outcome of compounding is to a very large extent unpredictable and unexplainable. That is, we cannot really explain why a particular quasi-idiomatic compound came to be associated with a particular idiomatic meaning. A good example illustrating this point is the quasi-idiom HEADACHE PILL. As a first approximation, it can be observed that its signified is rather similar to the signified of the quasi-idiom SUN HAT, mentioned above: a sun hat is a hat that is supposed to protect from the sun and a headache pill is a pill that is supposed to act as a pain reliever, i.e. as a pill that reduces the headache. At the same time, notice that the idiomatic meaning 'to reduce' is not the only logically possible idiomatic meaning which the compounding of the components *headache* and *pill* could have given rise to. Apart from the idiomatic meaning 'to reduce', *headache pill* could also have acquired the idiomatic meaning 'to cause': that is, a headache pill is a pill that causes the headache as a side-effect of the treatment of another condition. This quasi-idiomatic interpretation of *headache pill* is no less plausible than that of a pill that reduces the headache (Cf. Bauer 1979: 45-46, who discusses the semantic structure of the quasi-idiomatic compounds SLEEPING PILL, SEA-SICKNESS PILL, and ANTIHISTAMINE PILL).

5.6.3 Endocentric and exocentric compounding

Compounds are often classified into endocentric and exocentric compounds (or simply **endocentrics** and **exocentrics**). The distinction between these two categories relies heavily on the notion of the head of a compound, which, as we will see below, has both a semantic and a formal dimension. The unfortunate consequence of this is that one and the same compound can simultaneously qualify as an exocentric compound from a semantic point of view and as an endocentric compound from a formal point of view.

As regards the semantic dimension, recall the converted verb *to wife* and the affixed noun *reader*, which we discussed in the previous sections of this chapter. As we argued, complex signifieds like 'to wife' and 'a reader' can be represented as syntactic phrases (i.e. VPs, NPs, etc.) which are headed by a particular element. For example, the signified 'to wife' can be represented as the VP *to downplay a woman's career accomplishments in favor of her abilities as wife and mother*, whose head is the verb *downplay*. This fact explains why the quasi-idiomatization of the input nominal signified 'a wife' resulted in the word class change of the nominal input lex *a wife*. Similarly, the signified 'a reader' can be represented as the NP *a performer of the action of reading*, whose head is the noun *performer*. This fact explains why the affixation of the nominal input lex *read* by means of the suffix *-er* gave rise to an output lex which is a member of a different word class than the corresponding verbal input lex *read*. Comparing the two examples, one can notice that while in the case of *reader*, the head meaning 'performer' is part of the signified of the suffix *-er*, the head meaning 'to downplay' of *to wife* is an idiomatic meaning which is not part of the signified of the component *wife*. Complex lexemes like READER whose head meanings are inherent in one of their components' signifieds are endocentrics, while lexemes like TO WIFE whose head meanings are idiomatic meanings that are not part of their components' signifieds are exocentrics.

Returning to compounds, it appears that the endocentric–exocentric distinction neatly corresponds to the distinction between the two types of quasi-idiomatic compounds that was elaborated on in the previous part of this section. That is, while quasi-idiomatic compounds of the *information fatigue*-type always qualify as endocentric compounds – for example, the quasi-idiom INFORMATION FATIGUE is headed by the meaning 'fatigue', inherent in the component *fatigue*; the quasi-idiom EURO NOTE is headed by the meaning 'note', inherent in the component *note*; the quasi-idiom TO EGO-SURF is headed by the meaning 'to surf', inherent in the component *surf*; etc. – quasi-idiomatic compounds of the *drum and bass*-type are always exocentric compounds: e.g. the quasi-idiom DRUM AND BASS is headed by the idiomatic meaning 'music style', which is not inherent in either *drum* or *bass*; the quasi-idiom STAR 69 is headed by the idiomatic meaning 'service', which is not inherent in either *star* or *69*; the quasi-idiom NO-MATES is headed by the meaning 'person', which is not inherent in either *No* or *Mates*; etc.

At the same time, observe that the endocentric–exocentric distinction is broader than that between quasi-idiomatic compounds of the *information fatigue*- and *drum and bass*-types. In contrast to the latter, the former distinction can also be applied to semi- and fully-idiomatic compounds. Consider, for example, the semi-idiomatic compound lexeme DADROCK. Despite the fact that this semi-idiom does not qualify as a quasi-idiomatic compound of the *information fatigue*-type, it does qualify as an endocentric compound: its signified can be represented as the NP *rock that appeals to an older generation*, whose head element is the noun *rock*. Since the head meaning 'rock' is inherent in the component *rock*, we are justified in claiming that DADROCK is an endocentric compound. By contrast, the fully-idiomatic compound lexeme CARPET MUNCHER is clearly an exocentric compound. Its head meaning 'lesbian' is inherent in neither *carpet* nor *muncher*: like quasi-idiomatic compounds of the *drum and bass*-type, the components *carpet* and *muncher* only name a feature that can be metaphorically conceived of as an important characteristic of a lesbian.

In addition to endocentric and exocentric compounds, many authors also introduce the category **'copulative compound'** as a distinct compound type. Copulative compounds are said to be different from endocentric compounds in that both of their components are equally important from a semantic point of view. An often cited example is the compound FIGHTER-BOMBER 'an aircraft that combines the functions of a fighter and a bomber' (OED), whose components *fighter* and *bomber* denote equally important characteristics of fighter-bomber aircrafts: these can be used as both fighters and bombers. Similarly, the already mentioned compound POLITICIAN-TYCOON can also be regarded as a copulative compound because its lex can be used to refer to people who are both tycoons and politicians.

This textbook rejects the category of a copulative compound. What other studies regard as copulative compounds are compound lexemes which are in no essential respect different from exocentric compounds of the *drum and bass*-type. That is, the signified 'fighter-bomber' can be represented as the NP *an aircraft that can be used as both a fighter aircraft and a bomber aircraft*. This NP is headed by the meaning 'aircraft', which is not inherent in either *fighter* or *bomber*: in the compound under analysis, the components *fighter* and *bomber* only name the two equally important characteristics of a fighter-bomber aircraft distinguishing it from other military aircrafts. Similarly, the signified 'politician-tycoon' can be represented as the NP *a person who is both a politician and a tycoon*. This NP is headed by the meaning 'person', which is not inherent in either *politician* or *tycoon*: in the compound under analysis, the components *politician* and *tycoon* only name the two equally important characteristics of politician-tycoons distinguishing them from other people. Given these semantic structures, both *fighter-bomber* and *politician-tycoon* can be analyzed as exocentric compounds: they are headed by meanings which are not inherent in either of their components' signifieds.

Now, let us proceed to the endocentric–exocentric distinction from a formal point of view. Recall that analyzing the morphemic structure of the semantically opaque verb *understand*, we said that the formal head of the lex of a complex lexeme is that component which determines its inflectional marking. Thus even without considering the signified 'to understand', we can conclude that this verb is headed by the component *stand*: there can only be *he understand<u>s</u>* but not **he under<u>s</u>stand*; likewise, there can only be *he understood*, not **he understand<u>ed</u>*, let alone **he under<u>ed</u>stand*. Similarly, even without considering the signified 'fighter-bomber', we can easily analyze this compound as a formally endocentric compound headed by the component *bomber*: there can only be *two fighter-*

bombers but not **two fighters-bomber* or **two fighters-bombers*. Similarly, there can only be *two politician-tycoons* but not **two politicians-tycoon* or **two politicians-tycoons*.

A well-known characteristic of English compounds is their formal right-headedness, i.e. the righthand member receives the inflectional marking and thus functions as the formal head of a compound. This characteristic enables us to distinguish compounds from other idiomatic combinations involving two or more roots. For example, the fully-idiomatic VP *kick the bucket* 'to die' cannot be regarded as a compound because it is headed by the lefthand member *kick* but not by the righthand member *bucket*: we can only say *He kicked the bucket last night* but not **He kick the bucketed last night*. Similarly, so-called **phrasal verbs** such as, for example, the fully-idiomatic *come across* 'to find by chance' cannot be regarded as compounds because they are formally headed by their lefthand components: we can only say *He came across her diary* but not **He come acrossed her diary*.

The general right-headedness of compounds in English allows us to discard the endocentric–exocentric distinction, as far as the formal perspective is concerned. Indeed, if the righthand member always functions as the formal head of a compound, then all compounds can be regarded as endocentric compounds. Even if there were left-headed compounds (e.g. **two fightersbomber*), we would still be justified in regarding them as endocentric compounds, i.e. if *fighter-bomber* were headed by the lefthand component *fighter* rather than by the righthand component *bomber*, it would nevertheless qualify as an endocentric compound because in this putative compound the head position would still be filled by one of its overt components, the lefthand member *fighter*.

Finally, let us consider the possibility of **formal double-headedness**. As the term suggests, a double-headed compound is a compound that has two formal heads. As an illustration, let us consider the fully-idiomatic verbal compound TO DRAG AND DROP 'to move or copy (an image, icon, text, etc.) from one part of a display screen to another using a mouse or similar device'. According to the OED, the past tense form of this verb can be both *dragged and dropped* and *drag and dropped*. In the latter case, we are dealing with an endocentric compound which is headed by the righthand member *drop*. By contrast, *dragged and dropped* can be regarded as a **formally copulative compound**, i.e. a compound which is headed by both *drag* and *drop*. Alternatively, we can argue that the lexeme TO DRAG AND DROP is realized by the following two allolexes: 1) the endocentric compound lex *drag and drop*, whose past tense form is *drag and dropped* and 2) the copulative phrase *drag and drop*, whose past tense form is *dragged* and *dropped*.

In summary, from a semantic point of view, English compounds can be classified into endocentric and exocentric compounds. The former are compounds whose head meanings are inherent in one of their components' signifieds; the latter are compounds whose head meanings are idiomatic meanings non-inherent in their components' signifieds. Endocentric compounds include all quasi-idiomatic compounds of the *information fatigue*-type and semi-idiomatic compounds like *dadrock*. Exocentric compounds include all quasi-idiomatic compounds of the *drum and bass*-type (including bahuvrihis like *skinhead*) and fully-idiomatic compounds like *carpet muncher*. There are no copulative compounds: compounds like *fighter-bomber* and *politician-tycoon* which are traditionally classed as copulative compounds belong in the category of exocentric compounds.

From a formal point of view, English compounds can be classified into endocentric and copulative compounds. The former are compounds like *fighter-bomber* which are formally headed by their righthand components; the latter are compounds like *to drag and drop* which are formally headed by both their components. The overwhelming majority of English compounds are formally endocentric. (Forms like *to drag and drop* can also be analyzed as copulative phrases rather than as copulative compounds). There are no formally exocentric compounds.

Since the term 'head' has both a formal and a semantic dimension, one and the same compound can simultaneously qualify as e.g. an exocentric compound from a semantic point of view and as an endocentric compound from a formal point of view. E.g. *fighter-bomber* is a semantically exocentric, but formally endocentric compound.

5.6.4 Compounding from a formal point of view

As regards the formal side of compounding, the most important question is which input lexes can be combined into output lexes realizing new compound lexemes. Considering the compounds that were discussed in the previous parts of this section, we can already make several generalizations. First of all, nominal compounds can represent combinations of:

1. **two input nouns**: e.g. FIGHTER-BOMBER
2. **one input adjective and one input noun**: e.g. BLUE STATE
3. **one input verb and one input noun**: e.g. PLAYGROUND

As for verbal compounds, consider the following recent formations:

- TO CROWD-SURF 'to engage in crowd-surfing' (OED / 1991)

- TO DUMBSIZE 'to dismiss (staff) in excessive numbers or without regard to organizational function, with the result that work can no longer be carried out effectively' (OED / 1993)

- TO MWAH-MWAH 'to kiss in exaggerated fashion, especially on the cheek; to give air kisses' (OED / 1993)

- TO COPYPASTE '[to] copy[…] the contents of a document or a program to be added to another document' (Urban Dictionary / 2004)

TO CROWD-SURF is, like the previously mentioned TO EGO-SURF, a **noun + verb compound**: its input lexes are the noun *crowd* and the verb *to surf*.

TO DUMBSIZE is an **adjective + verb compound**: its input lexes are the adjective *dumb* and the verb *to size*.

TO MWAH-MWAH is a reduplicative **interjection + interjection compound**: its input lex is the interjection *mwah* 'representing the sound of a kiss, deliberately exaggerated to convey superficiality or pretentiousness' (OED).

Finally, TO COPYPASTE is a **verb + verb compound**: its input lexes are the verbs *to copy* and *to paste*.

Adjectival compound lexemes usually exhibit either 1) the **adverb + adjective** pattern, exemplified by the already mentioned adjectival compounds ALWAYS-ON 'always online' and DOWN-LOW 'secret, quiet' – both are products of combining the input adverbial lexes *always* and *down* and the adjectival lexes *on* and *low* – or 2) the **adjective + adjective** pattern. With regard to the latter, consider the adjectival compound BI-CURIOUS 'of a heterosexual person: interested in experiencing an (especially first) sexual encounter or relationship with a person of the same sex' (OED / 1990). Its input lexes are the adjectival *bi* (which is a colloquial abbreviation of *bisexual*) and the adjectival *curious*.

Finally, let us briefly dwell on the following special types of compounds:

- **reduplicative compounds**
- ***thing*-compounds**
- **neo-classical compounds**

Reduplicative compounds are compounds like the above mentioned TO MWAH-MWAH whose lexes are formed via reduplication of one and the same input lex: *to mwah-mwah = mwah + mwah*. Other examples include:

- TO NYAH-NYAH 'to behave in a childishly supercilious or derisive manner towards someone; to taunt someone' (OED / 1986)

- PUM-PUM 'the female external genitals, the vagina […]' (OED / 1983)

- the interjection BOOM-BOOM 'used (as a following tag or as a response) to draw attention to a joke or pun, especially one the speaker or writer regards as weak, obvious, or labored' (OED / 1972)

- the adjective GO-GO 'fashionable, 'swinging', 'fabulous', unrestrained; (of funds on the stock exchange) speculative' (OED / 1962)

- the adverb NOW-NOW 'in the immediate future, in a moment; very soon' (OED / 1948)

Sometimes reduplicative compounding is accompanied by apophony. For example:

- the adjective EASY-PEASY 'in childish or children's use (especially as interjection): extremely easy, very simple' (OED / 1976)

- the noun RUMPY-PUMPY 'sexual intercourse' (OED / 1968)

- the interjection OKEY-DOKEY 'OK' (OED / 1932)

- the adjective FLIPPY-FLOPPY 'inclined to flop, having a tendency to flop about' (OED / 1858).

Thing-compounds are noun + noun compounds like *bodybuilding thing* in (85) and *jet thing* in (86).

(85)　*I don't like that <u>bodybuilding thing</u> where they got no neck at all, and then every vein pops out* (COCA)
(86)　*And it is really fantastic to have your own jet, and anybody who says it isn't is lying to you. That <u>jet thing</u> is really good* (COCA)

A notable peculiarity of *thing*-compounds is that they never become fully-established lexemes and are therefore highly context-dependent (Lieber 2009: 365). Thus it is the following part of the sentence *where they got no neck at all, and then every vein pops out* that enables us to understand that *thing* of *bodybuilding thing* stands for the particular outcome of bodybuilding. And it is the preceding sentence *and it is really fantastic to have your own jet, and anybody who says it isn't is lying to you* that enables us to understand that *thing* of *jet thing* stands for the state of possessing a jet.

Finally, neo-classical compounds are compounds that come into existence via combining roots of Greek and Latin origin (either with each other or with native roots). Consider, for example, the lexeme SARCOPENIA 'loss of skeletal muscle mass as a result of ageing' (OED / 1991). According to the OED, the lex of this lexeme is a product of combining the following roots of Greek origin: *sarco-* (σαρκ-, σάρξ meaning 'flesh') and *-penia* (Greek πενία meaning 'poverty, need'). However, despite the Greek origin of both *sarco-* and *-penia*, the lexeme *sarcopenia* is not an instance of borrowing: the compounding of the forms *sarco-* and *-penia* took place in English, not in Greek. Neoclassical compounds like *sarcopenia* usually consist of bound roots that never occur in isolation. (Given the general obligatoriness of roots, we cannot regard *sarcopenia* as a combination of two affixes. Either *sarco-* or *-penia* must be regarded as a root. The lex *sarcopenia* can thus be analyzed as either a compound segmentable into the two bound roots *sarco-* and *-penia* or as an affixed word segmentable into either the bound root *sarco-* and the suffix *-penia* or into the prefix *sarco-* and the bound root *-penia*. Both analyses are equally plausible here.)

5.6.5 Compounds and phrases

One of the most controversial theoretical issues regarding compounding in English is the question of how a compound can be distinguished from a phrase. In 5.6.3 we argued that fully-idiomatic VPs like *kick the bucket* and fully-idiomatic phrasal verbs like *come across* do not qualify as compounds because they are headed by their lefthand members *kick* and *come*. Compounds, by contrast, are right-headed: whereas the past tense forms of *kick the bucket* and *come across* are *kicked the bucket* and *came across*, the past tense form of the compound *ego-surf* is *ego-surfed*, not **egoed-surf*. Unfortunately, the right-headedness criterion is helpful only for distinguishing verbal compounds from idiomatic VPs and phrasal verbs. However, this criterion does not allow us to distinguish nominal compounds from idiomatic NPs. As an illustration, recall the compounds HOT DESK /ˈhɒt dɛsk/ 'a shared office desk or workstation, occupied on a temporary, ad hoc basis or part-time basis, and not allocated permanently to an individual' (OED / 1990) and BLUE STATE /ˈbluː steɪt/ 'a state (projected to be) won by the Democratic candidate in a presidential election' (OED / 2000), which were mentioned in 5.6.1. How do we know that the lexes realizing these lexemes are indeed semi-idiomatic adjective + noun compounds rather than semi-idiomatic adjective + noun NPs? To answer this question, we cannot resort to the right-headedness criterion because both adjective + noun compounds and adjective + noun NPs are right-headed: in both cases the plural marking occurs on the righthand head noun; e.g. *blackboards* and *black boards*.

Of little help also are the spelling criterion and the meaning criterion that are sometimes mentioned in connection with the compound–phrase distinction. As regards spelling, recall that orthographic systems are artificial systems, which appeared much later than spoken speech. Accordingly, the separation of *hot* and *blue* from *desk* and *state* by means of a blank space cannot be seen as a corroboration of the phrasal status of *hot desk* and *blue state*. As regards the meaning criterion that states that adjective + noun compounds are sometimes more idiomatic than corresponding homonymous phrases – cf. e.g. the semi-idiomatic signified of the compound *a blackboard* and the non-idiomatic signified of the homonymous NP *a black board* – observe that semi-idiomatic NPs do occur as well. For example, just like the compound *blackboard*, the adjective + noun NP *black coffee* has a semi-idiomatic signified that contains the signified of the component *coffee* but not the signified of the component *black*: *black coffee* does not mean 'black coffee' but 'coffee without milk or cream' (Mel'čuk 1995: 182). Does it follow from this that *black coffee* is a compound as well?

The only reason why the lexes realizing the semi-idiomatic lexemes HOT DESK and BLUE STATE can be regarded as compounds is their leftward stress: /ˈhɒt dɛsk/ and /ˈbluː steɪt/. The point here is that while the primary stress of an NP falls on the righthand element (e.g. *black cóffee*, not *bláck coffee*), compounds are stressed on their lefthand members. Accordingly, since in both *hot desk* and *blue state*, the stress falls on the lefthand members *hot* and *blue*, we can claim that both these combinations represent semi-idiomatic compounds rather than semi-idiomatic phrases. The same can be said about the lexes realizing the following lexemes:

- NEW JILL /ˈnjuː ˌdʒɪl/ 'new jack swing as performed by women' (OED / 1990)

- RIOT GIRL /ˈrʌɪət ɡəːl/ 'a member or follower of any of several loosely affiliated, mainly American, feminist rock or punk groups of the early 1990s' (OED / 1991)

- UK GARAGE /ˌjuːˈkeɪ ˌɡɑrɑː(d)ʒ/ 'a form of garage music [...] originating in the United Kingdom, retaining the emphasis on vocals but characterized by a syncopated rhythm track influenced by drum and bass' (OED / 1992)

Since in *new jill*, *riot girl*, and *UK garage*, the primary stress falls on the lefthand members *new*, *riot*, and *UK*, we are justified in regarding these forms as compounds.

Now, consider the stress pattern of the lexes realizing the following five lexemes:

- MARTIAL ARTISTRY /ˌmɑːʃl ˈɑːtɪstri/ 'achievement or skill in a martial art' (OED / 1990)

- ETHNIC CLEANSING /ˌɛθnɪk ˈklɛnzɪŋ/ 'the purging, by mass expulsion or killing, of one ethnic or religious group by another, especially from an area of former cohabitation' (OED / 1991)

- NEW LAD /ˌnjuː ˈlad/ 'a (type of) young man who embraces sexist attitudes and the traditional male role as a reaction against the perceived effeminacy of the 'new man"' (OED / 1991)

- HOME ZONE /ˌhəʊm ˈzəʊn/ 'a residential area in which a variety of traffic-calming measures are employed to create a safer environment for pedestrians' (OED / 1992)

- FREE RUNNING /ˌfriː ˈrʌnɪŋ/ 'the discipline or activity of moving rapidly and freely over or around the obstacles presented by an (especially urban) environment by running, jumping, climbing' (OED / 2003)

In all these forms the primary stress falls on their righthand members *artistry*, *cleansing*, *lad*, *zone*, and *running*. Accordingly, the lexes *martial artistry*, *ethnic cleansing*, *new lad*, *home zone*, and *free running* are all phrases, not compounds.

Many authors are rather skeptical about the applicability of the stress criterion (e.g. Giegerich 2009: 184-185) for distinguishing compounds from phrases. The main reason for this is that the difference in stress often necessitates a different treatment of semantically related combinations. For example, while the combination *apple pie* /ˌapl ˈpaɪ/ qualifies as a phrase because in British English it is stressed on the righthand element *pie*, the combination *apple cake* /ˈapl ˌkeɪk / qualifies as a compound because its primary stress falls on the lefthand element *apple*. For many authors, this analysis is rather counter-intuitive, given the semantic similarities between the combinations *apple pie* and *apple cake*.

In spite of this and other similar cases, this textbook argues for the stress criterion. The distinction between a compound and a phrase is a formal rather than a semantic distinction. This means that if we do not want to discard this distinction (i.e. to regard compounds and phrases as manifestations of the same formal category), we must ignore the fact that *apple pie* and *apple cake* express similar meanings. The only thing that must be taken into account is the formal difference between these combinations. That is, since *apple pie* is stressed on the righthand element *pie*, it must be regarded as a phrase and since *apple cake* is stressed on the lefthand element *apple*, it must be regarded as a compound. Moreover, the combination *apple pie*, which in American English is stressed on the lefthand element *apple* /ˈæpəl ˌpaɪ/, must be treated differently than the same combination in British English: while the British *apple pie* is a phrase, the American *apple pie* is a compound, even though both express the same meaning.

Finally, let us briefly discuss the status of double-stressed combinations. Consider again the above mentioned combination *new lad* /ˌnjuː ˈlad/. While in British English this combination is characterized by the rightward primary stress and thus qualifies as a phrase, in American English both *new* and *lad* receive the primary stress: /ˈn(j)u ˈlæd/. What is the status of *new lad* in American English? Is it a compound or a phrase? Or do we need a separate category for double-stressed combinations like *new lad*?

According to Bloomfield (1973[1934]: 180), double-stressed combinations must be regarded as phrases rather than as compounds. One rationale for this is that combinations like *new lad* are no different from phrases like *ethnic cleansing* in that one of their primary stresses also falls on their righthand members. (The only difference is that combinations like *new lad* have two primary stresses.) Another rationale for the phrasal solution is that words are often said to be able to carry only one primary stress. That is, a combination of two free morphs like *new* and *lad* of *new lad* both of which receive the primary stress cannot be regarded as a compound simply because a compound is a word and hence can only have one primary stress.

To conclude: the compound–phrase distinction is a formal, not a semantic distinction. Accordingly, this distinction must be specified in terms of formal properties of noun + noun and adjective + noun compounds distinguishing them from corresponding phrasal noun + noun and adjective + noun combinations which lack these properties. Of the available criteria, the difference in stress seems to be the most promising criterion: combinations like *hot desk* and *blue state* which stress their lefthand elements are compounds, whereas combinations like *ethnic cleansing* and *free running* in which the primary stress falls on their righthand elements as well as double-stressed combinations like *new lad* in American English are phrases.

5.6.6 Productivity

Compounding is a very productive lexeme-building mechanism. As was pointed out in 5.6.1, of 587 new signifiers which, according to the OED, appeared in the English language between 1990 and 2011, 81 are instances of pure compounding (i.e. not pseudo-compounds like TO BABYSIT).

As in the case of conversion, the Etymology-sections of the OED entries do not contain the word *compound*. This means that the OED is not searchable for compounds the way it is searchable for e.g. apophonies and borrowings. The best means to study the synchronic productivity of compounding using the OED is thus to make the OED search for words belonging to a particular word class (e.g. verbs, nouns, adjectives) that were formed during a particular period of time (e.g. after 1990) and then manually classify the compounds found into instances of particular categories (e.g. noun + noun compounds, exocentric compounds, neo-classical compounds, etc.). For example, if you want to find out how many nominal compounds appeared in English after 1990, go to the OED Advanced Search page http://www.oed.com/advancedsearch and choose the options '1990-' in 'Date of entry' and 'Noun' in 'Part of speech'. The OED will then yield all nominal lexemes which appeared in the English language after 1990. Then carefully read the Etymology-section of each entry. If it contains information like 'Etymology < *drum* n.1 + *and* conj.1 + *bass* n' for *drum and bass* or 'Etymology < *no* adj. + the plural of *mate* n.2' for *No-Mates*, you can be justified in regarding the signifiers under analysis as products of noun-building compounding.

The OED is searchable for compound lexemes whose lexes were formed via reduplication of their input lexemes. Go to the above named OED Advanced Search page, enter 'reduplication' to the first row of the search mask from above, and choose the option 'Etymology' (near the search mask to which you will enter 'reduplication'). The OED will then yield all entries whose Etymology-sections contain the word *reduplication*. Some of them will be instances of reduplicative compounding like *to mwah-mwah* and *to nyah-nyah*.

5.7 Blending

Blending is similar to compounding in that it also produces new lexemes by combining the lexes of two or more input lexemes into the lex of a new complex lexeme. However, in contrast to compounding, blending is accompanied by the shortening of at least one of the input lexes. For example, the compounding of the input lex *flu* and *tsunami* into the blend *flunami* was accompanied by the shortening of one of its input lexes: *tsu̶nami* → *nami*. Similarly, the compounding of the input lexes *flexibility* and *security* into the blend *flexicurity* 'labor practices that give companies the flexibility to fire workers as needed and offer fired workers the security of government-backed benefits and retraining opportunities' (Word Spy / 1997) was accompanied by the shortening of both of its input lexes: *flexibi̶li̶ty̶* → *flexi* and *se̶curity* → *curity*.

Sometimes it may be rather difficult to determine which of the input lexes that gave rise to a new blended lexeme was subjected to shortening. Consider, for example, the blend *cheapuccino* 'an inexpensive, low-quality cappuccino, particularly one from a vending machine [...]' (Word Spy / 2002). Has the lex of this lexeme been created via the shortening of *cheap* into *chea* and *ca̶ppuccino* into *puccino* (i.e. *chea* + *puccino*) or only via the shortening of *ca̶ppuccino* into *uccino* (i.e. *cheap* + *uccino*)? Both analyses seem equally plausible here, so that the exact diachronic history of this blend cannot be established. However, the question of whether *cheapuccino* = *chea* + *puccino* or *cheap* + *uccino* does not seem to be an important theoretical question. The only thing that matters here is that the lex under analysis is indeed a product of blending of the input lexes *cheap* and *cappuccino*.

Like compounding, blending is an anisomorphic lexeme building-mechanism, which always produces output lexemes whose signifieds are not / not entirely representable in terms of their components' signifieds. As in the case of compounding, quasi-idiomatization is the default semantic outcome of blending. Thus of 20 blended signifiers which, according to the OED, appeared in the English language between 1990 and 2011, 17 can be analyzed as quasi-idioms. For example:

- EMOTICON (← ~~emot~~*ion* + *icon*) 'a representation of a facial expression formed by a short sequence of keyboard characters (usually to be viewed sideways) and used in electronic mail, etc., to convey the sender's feelings or intended tone' (OED / 1990)

- CLINTONOMICS (← ~~Bill~~ *Clinton* + ~~ec~~*onomics*) 'the economic policies of President Clinton' (OED / 1992)

- RACINO (← *race*~~track~~ + ~~c~~*asino*) 'a building complex or grounds having a racetrack and gambling facilities traditionally associated with casinos, such as slot machines, blackjack, roulette, etc.' (OED / 1995)

While the blended lexemes EMOTICON and CLINTONOMICS are quasi-idioms of the *information fatigue*-type – that is, an emoticon is a particular icon that has something to do with emotions and Clintonomics are particular economic policies that have something to do with Clinton – the blend RACINO exhibits the *drum and bass*-pattern: its blended components name two important characteristics of building complexes that came to be referred to as racinos: they contain both racetracks and casinos. The blends EMOTICON and CLINTONOMICS are thus semantically endocentric blends, whereas RACINO is a semantically exocentric blend.

Examples of semi-idiomatic blends include:

- SCREENAGER (← *screen* + ~~te~~*enager*) 'a young person (typically in his or her teens or twenties) who is at ease with and adept at using new technology and media, especially computers' (OED / 1994)

- MACHINIMA (← *mach*~~ine~~ + ~~c~~*inema* or *machine* + ~~c~~*inema*) 'the practice or technique of producing animated films using the graphics engine from a video game' (OED / 2000)

Both these blends involve a metonymic modification of their first components *screen* and *machine*: *screen* of *screenager* stands for a new technology which typically involves screens (e.g. computer screens) and *machine* of *machinima* stands for computers with the help of which such films are created.

The only example of a fully-idiomatic blend yielded by the OED is CRUNK (← *crazy* + *drunk* or *crazy* + *drunk*) 'exciting or fun; (of a person) extremely energized or excited, especially as a result of listening to (usually hip-hop or rap) music' (OED / 1995).

5.7.1 Productivity

Blending is often described as a minor lexeme-building mechanism in English. Indeed, while e.g. compounding has recently given rise to 81 new lexemes, blending has created only 20 new lexemes. At the same time, note that the Word Spy database lists 189 lexemes whose lexes have recently come into existence via blending. For example:

- MANUFACTROVERSY (←*manufactured* and *controversy*) 'a contrived or non-existent controversy, manufactured by political ideologues or interest groups who use deception and specious arguments to make their case' (Word Spy / 2008)

- CELEBUTARD (←*celebutante* and *retard* or *celebutard* and *retard*) 'a celebrity who is or is perceived to be unintelligent' (Word Spy / 2006)

- RENOVICTION (←*renovation* and *eviction* or *renovation* and *eviction*) 'the mass eviction of an apartment building's tenants because the building's owner plans a large renovation' (Word Spy / 2008)

- CHURNALISM (←*churn* and *journalism*) 'journalism that churns out articles based on wire stories and press releases, rather than original reporting' (Word Spy / 2001)

- SPIME (←*space* and *time*) 'a theoretical object that can be tracked precisely in space and time over the lifetime of the object' (Word Spy / 2004)

- LUSER (← *loser* and *user*) 'a person who doesn't have the faintest idea what they're doing and who, more importantly, refuses to do anything about it' (Word Spy / 1990)

- BLOGEBRITY (←*blog* and *celebrity*) 'a famous or popular blogger' (Word Spy / 2005)

- WOMENOMICS (← *women* and *economics*) 'the theory that women play a primary role in economic growth' (Word Spy / 1995)

- WARMEDY (← *warm* and *comedy*) 'a comedy that features warm-hearted, family-oriented content' (Word Spy)

According to the author of the Word Spy database, "blends are the engines that drive much of neology". In other words, blending is currently one of the most productive lexeme-building mechanisms in English.

To study the productivity of blending using the OED, go to the OED Advanced Search page http://www.oed.com/advancedsearch, enter 'blend' to the search mask of the first row from above, and choose the option 'Etymology' (near the search mask to which you will enter 'blend'). The OED will then yield all entries whose Etymology-sections contain the word *blend*. Some of them will be instances of blending like the above mentioned *emoticon*, *racino*, *screenager*, etc.

To study the productivity of blending using Word Spy, go to the Advanced Search page http://wordspy.com/search.asp, enter 'blend' to the search mask, and choose the option 'Anywhere'. Word Spy will then find all items whose descriptions contain the word *blend*. Since blended signifiers are almost always described by the database as instances of blending, the majority of the results yielded by Word Spy will indeed be recently created blended lexemes like *manufactroversy*, *spime*, *luser*, etc.

5.8 Idiomatization of phrases and sentences

Like compounding and blending, phrasal idiomatization is an anisomorphic lexeme-building mechanism that produces new lexemes via combining the lexes of more than one input lexeme into the lex of a new complex lexeme. In fact, as we established in 5.6.5, idiomatic phrases (especially idiomatic NPs such as, for example, *new lad*) are sometimes hardly distinguishable from idiomatic compounds.

As we said in 3.2.5, idiomatic phrases and sentences exhibit the same degrees of idiomaticity as idiomatic words. That is, there are not only quasi-idiomatic words like *football* but also quasi-idiomatic phrases like the VP *start a family* and quasi-idiomatic sentences like the pick-up line *Would you like to have morning coffee with me?*, whose signifieds contain not only their components' signifieds but also some unpredictable idiomatic signifieds.

Similarly, there are not only semi-idiomatic words like *blackboard* but also semi-idiomatic phrases like the VP *answer the door* and the pick-up line *Didn't I see you on the cover of Vogue?*, whose signifieds contain only some of their components' signifieds. Finally, there are not only fully-idiomatic words like *boyfriend* but also fully-idiomatic phrases like the VP *kick the bucket* and the proverb *A bird in the hand is worth two in the bush*, whose signifieds contain none of their components' signifieds.

5.8.1 Productivity

Like compounding and blending, phrasal idiomatization is a fairly productive lexeme-building mechanism in English. Thus the Idioms-section of the Word Spy database lists 18 recently formed idiomatic phrases. Below are some of the examples:

- TO STARVE THE BEAST 'to cut taxes with the intent of using the reduced revenue as an excuse to drastically reduce the size and number of services offered by a government' (Word Spy / 1981)

- TO PUT SKIN IN THE GAME 'to take an active interest in a company or undertaking by making a significant investment or financial commitment' (Word Spy / 1993)

- TO PAINT THE TAPE 'to increase the price of the stock by using unscrupulous methods (such as breaking up a large stock purchase into multiple small purchases to give the illusion of a buying frenzy)' (Word Spy / 2000)

- DOG THAT CAUGHT THE CAR 'a person who has reached their goal but doesn't know what to do next' (Word Spy / 1985)

- DOG WATCHING TV 'a person who is viewing or working with something without understanding what it is or what it does' (Word Spy / 1997)

Fully-idiomatic phrases and sentences have traditionally been the focus of attention of phraseology rather than morphology. The explanation for this is that morphology, as a branch of linguistics, is supposed to be concerned with words, not with phrases and sentences. Similarly, lexicographers tend to regard phrasal idiomatization as a special lexeme-building mechanism distinct from conversion, affixation, compounding, and the like. As a consequence, a number of established idiomatic phrases can only be found in specialized idiom dictionaries, but not in 'word' dictionaries like the OED and the MWO.

One specialized dictionary that can be recommended for the study of idiomatic phrases is Ayto et al's *Oxford Dictionary of English Idioms* (2009), which contains more than 6000 entries. An entry in this dictionary is usually a noun or a verb that occurs in multiple idiomatic phrases: e.g. the entry for the noun *door* lists such phrases as:

- *at death's door*
- *beat a path to someone's door*
- *blow the doors off*
- *by the back door*

Like the OED, this dictionary can be used online. However, to gain access to the dictionary, you need to subscribe to Oxford Reference Online Premium services.

Apart from this, there are a number of freely-available databases collecting English idiomatic phrases. For instance, the already mentioned Phrase Finder (http://www.phrases.org.uk/) is a good resource for studying the etymologies of idiomatic phrases.

5.9 Back-formation

Back-formation is the removal of a derivational affix (or a part of an input lex perceived as a derivational affix) from the lex of an input lexeme accompanied by quasi-idiomatization of the signified of an input lexeme. For example, the removal of *-er* from the input lex *Taser* was accompanied by the quasi-idiomatization of the input signified 'Taser': while *a Taser* means 'a Taser', *to tase* came to mean 'to incapacitate using a Taser'. Likewise, the removal of *-y* from the input lex *skeevy* was accompanied by the quasi-idiomatization of the input signified 'skeevy': while *skeevy* means 'skeevy', *a skeeve* came to mean 'a skeevy person'. Similar examples include:

- TO CARJACK (←*carjacking*) 'to steal or to commandeer an occupied car by threatening the driver with violence' (OED / 1991)

- TACK (←*tacky*) 'that which is 'tacky' or cheap and shabby; shoddy or gaudy material; rubbish, junk' (OED / 1986)

- A DITZ (←*ditzy*) 'a person who is 'ditzy', scatterbrained, or cute' (OED / 1984)

- TO DIVISIONALIZE (←*divisionalization*) 'to organize (a company, etc.) on a divisional basis' (OED / 1982)

Lexeme-building mechanisms

- A SHONK (*shonky*) 'one engaged in irregular or illegal business activities; a 'shark'' (OED / 1981)

- TO EXFILTRATE (←*exfiltration*)' to withdraw (troops, spies, etc.) from a dangerous position' (OED / 1980)

- TO CHEMOTAX (←*chemotaxis*) 'to exhibit chemotaxis; to move in response to certain chemical substances' (OED / 1979)

- TO INCENT (←*incentive*) 'to provide (a person) with an incentive; to encourage, incite, inspire' (OED / 1977)

- A GIGAFLOP (←*gigaflops*) 'a unit of computing speed equal to 1000 megaflops' (OED / 1976)

As these examples illustrate, back-formation typically gives rise to verbal and nominal lexemes. As for adjectives, it appears that back-formation is no longer used for producing new adjectival lexemes: according to the OED, the last adjectival back-derivative is INTERTEXTUAL (← *intertextuality*) 'denoting literary criticism which considers a text in the light of its relation to other texts; also used of texts so considered' (OED / 1973). Since the early 1970s not a single adjectival lexeme has been formed by means of back-formation.

As can be inferred from what was said in 4.2.4, the most important theoretical question raised by back-formation is when English speakers 'incorrectly' reanalyze instances of back-formation as input lexemes. That is, while *to tase* still means 'to incapacitate using a Taser' and, accordingly, can be regarded as an instance of back-formation not only from a diachronic but also from a synchronic point of view, *to babysit* and *to proofread* no longer mean 'to do the job of a babysitter' and 'to do the job of a proofreader': as we argued in 4.2.4, the semantic relation holding between 'to babysit' / 'to proofread' and 'a babysitter' / 'a proofreader' is in Present-day English no different from the semantic relation holding between the signifies 'to blog' and 'a blogger'. Accordingly, from a synchronic point of view, the signifieds 'to babysit' / 'to proofread' must be regarded as input signifieds for 'a babysitter' / 'a proofreader'. In which respects is TO TASE different from TO BABYSIT and TO PROOFREAD?

The first obvious difference is that while TO TASE is a relatively recent formation – it has been recorded in English since 1991 – the earliest citation of TO BABYSIT dates 1947 and the earliest citation of TO PROOFREAD dates 1845. Given these dates, we can conjecture that speakers of English have already forgotten the true etymologies of the lexemes TO BABYSIT and TO PROOFREAD, but they have not forgotten the true etymology of TO TASE.

Another important difference is that in contrast to the lex *to tase*, the lexes *to babysit* and *to proofread* do not fulfill the **additional naming requirement**. (Dobrovol'skij and Piirainen 2005: 18). That is, *to babysit* is a **primary lex** (i.e. the most basic signifier) expressing the signified 'to take care of a baby during the temporary absence of the parents'. Likewise, *to proofread* is the most basic signifier expressing the signified 'to read a proof, identify mistakes in it, and make the necessary corrections'. By contrast, *to tase* seems to be an additional expression for the signified 'to incapacitate a person using a Taser'. Usually this signified is expressed by the euphemistic VP *to temporarily incapacitate a person*. The point here is that in the case of primary expressions like *to babysit* and *to proofread*, language users often 'get rid' of the diachronic memory: there is absolutely no need to remember that e.g. *to babysit* and *to proofread* came into existence via back-formation of the corresponding nouns *babysitter* and *proofreader*. In contrast, in the case of additional expressions like *to tase*, the true etymology usually provides a motivating link explaining why a particular signifier can be used as an additional signifier expressing a particular signified. That is, for example, the fact that *to tase* came into existence via back-formation of *Taser* provides a synchronic motivation for the possibility of the use of that signifier for expressing the signified 'to incapacitate a person using an electroshock weapon like a Taser'. If it were not for this motivating link between the signifier *to tase* and the signified 'to incapacitate somebody using a Taser', the lexeme TO TASE would be synchronically opaque. (For a more detailed discussion of the additional naming requirement, see Tokar 2009: 10-11).

5.9.1 Productivity

At present, back-formation does not seem to be a productive lexeme-building mechanism in English. According to the OED, the most recent instances of back-formation – the verbal lexemes TO TASE and TO CARJACK – appeared in 1991. Since then there have been no new back-formations.

To study the history of this lexeme-building mechanism, go to the OED Advanced Search page at http://www.oed.com/advancedsearch, enter 'back-formation' to the search mask of the first row from above, and choose the option 'Etymology' (near the search mask to which you will enter 'back-formation'). The OED will then yield all entries whose Etymology-sections contain the word *back-formation*. Some of them will be signifiers like *to tase* and *to carjack* which came into existence via back-formation. You can refine your search by choosing the option 'Noun' and entering e.g. '1900-' to 'Date of entry'. The OED will then yield all nominal signifiers which came into existence via back-formation between 1900 and 2011.

5.10 Exercises

1. Make sure you can explain each of the key terms printed in boldface (ideally, using your own examples).

2. Which of the following statements are true?

a) Metonymy and metaphor are two different mechanisms of semantic change.
b) Morphological conversion is no longer a productive lexeme-building mechanism in English.
c) Arbitrary formation is the most productive lexeme-building mechanism in Present-day English.
d) English has extensively borrowed from other languages.
e) The addition of a derivational suffix often produces an output lexeme whose lex is a member of a different word class than the lex of a corresponding input lexeme.
f) Compounds in English are left-headed.
g) The head meaning of an exocentric compound is not inherent in its components' signifieds.
h) Full-idiomatization is the default semantic outcome of compounding.
i) Adjective + noun compounds have the leftward stress and are in this respect different from phrases.
j) Blending is now a very productive lexeme-building mechanism in English.

3. Using the OED, establish how many verbs which appeared in English in 1989 are products of morphological conversion.

4. Using the OED, establish which of the following languages – German or Spanish – played a greater role in the enlargement of the English vocabulary in the 20th century.

5. Using the OED, establish whether lexeme-manufacturing was a productive lexeme-building mechanism in English between the years 1300 and 1350.

6. Using the OED, establish how many verbs which appeared in English between the years 1900 and 1950 can be regarded as instances of apophony.

7. Using the OED, establish how many nouns which appeared in English in 1990 are products of affixation by means of the suffix *-er*.

8. Using the OED, establish whether blending gave rise to more new lexemes between the years 1600-1649 or between the years 1900-1949.

9. Using the OED, establish how many nominal lexemes were formed in English via back-formation between the years 1700-1749.

5.11 Further reading

A classic monograph dealing with lexeme-formation in English is Marchand (1969). A more recent discussion of the most important theoretical issues pertaining to English word-formation can be found in the recent *Handbook of Word-Formation* (Štekauer and Lieber 2005) and *The Oxford Handbook of Compounding* (Lieber and Štekauer 2009b). A recent study of the directionality of conversion in English is Balteiro (2007). A recent study of the semantics of compounding in English is Benczes (2006).

Due to space limitations, this chapter does not discuss lexeme-formation in Old and Middle English. The reader is referred to the articles Kastovsky (1992) and Burnley (1992) from *The Cambridge History of the English Language* (Volumes I and II). The former discusses lexeme-formation in Old English; the latter deals with the Middle English period.

6 Inflectional morphology

Having discussed both the lex- and lexeme-building mechanisms, we can finally proceed to wordform-building mechanisms, i.e. mechanisms like inflectional affixation producing output allolexes which express different grammatical meanings than corresponding input lexes (e.g. *talked* and *talk*, *books* and *book*, *prettier* and *pretty*). The chapter has the following structure. Section 6.1 provides a more precise definition of the term 'grammatical category', which was already introduced in 1.1. Section 6.2 classifies grammatical categories into syntactic and semantic grammemes. Section 6.3 introduces all wordform-building mechanisms with the help of which speakers of English create wordforms like *talked*, *books*, *prettier*, etc. Finally, Sections 6.4 and 6.5 dwell on the most important theoretical issues pertaining to both syntactic and semantic grammemes in English.

6.1 Grammatical category

A grammatical category is the set of **mutually exclusive** grammatical meanings such as, for example, 'the singular number' and 'the plural number' (forming the grammatical category NUMBER) or 'the present tense' and 'the past tense' (forming the grammatical category TENSE).

To explain what is meant by the 'mutual exclusiveness' of grammatical meanings, let us recall what we said about the contrast between the past tense meaning inherent in the inflectional suffix *-ed* of e.g. *talked, walked, worked* and the past time meaning inherent in the derivational prefix *ex-* of e.g. *ex-ambassador, ex-boyfriend, ex-president*: while the latter is an optional lexical meaning that is expressed only when we want to specifically refer to people who are former ambassadors, boyfriends, and presidents, the former is an obligatory grammatical meaning, i.e. a meaning which cannot be unexpressed. This characterization of grammatical meanings cannot but raise the following objection: if the past tense meaning inherent in *-ed* were indeed an obligatory meaning, than all verbal lexemes would have only past tense wordforms (*talked, walked, worked*). Similarly, if the meaning 'the plural number' were an obligatory nominal meaning, than all nominal lexemes would consist of only plural wordforms like *books, chairs, tables*. But as we know, verbal lexemes have present tense wordforms (*talks, walks, works*) and nominal lexemes have singular wordforms (*book, chair, table*).

As Plungian (2000: 107) points out, obligatoriness is not a characteristic of a particular grammatical meaning such as 'the past tense' or 'the plural number' but of a set of mutually exclusive grammatical meanings of which that particular meaning is a member. For example, the meaning 'the past tense' is a member of the set of the mutually exclusive meanings 'the present tense' / 'the past tense'; the meaning 'the plural number' is a member of the set of the mutually exclusive meanings 'the singular number' / 'the plural number'; the meaning 'the comparative degree of comparison' is a member of the set of the mutually exclusive meanings 'the positive degree of comparison' / 'the comparative degree of comparison' / 'the superlative degree of comparison'; etc.

The mutual exclusiveness of grammatical meanings means that in no case can two meanings of the set in question be simultaneously expressed by one and the same wordform: e.g. there can be no verbal wordforms that express both the meanings 'the present tense' and 'the past tense' and there can be no nominal wordforms that express both the meanings 'the singular number' and 'the plural number'. But one of the meanings of the set must always be expressed by a given wordform. That is, a wordform of a verbal lexeme must express either the meaning 'the present tense' or 'the past tense'; a wordform of a nominal lexeme must express either the meaning 'the singular number' or 'the plural number'; a wordform of an adjectival lexeme must express either the meaning 'the positive degree of comparison' or 'the comparative degree of comparison' or 'the superlative degree of comparison'; etc.

Meanings that are members of such mutually exclusive sets of grammatical meanings form a grammatical category (or a **grammeme**). For example, as said above, the mutually exclusive verbal meanings 'the present tense' and 'the past tense' form the grammatical category TENSE; the mutually exclusive nominal meanings 'the singular number' and 'the plural number' form the grammatical category NUMBER; the mutually exclusive adjectival meanings 'the positive degree of comparison', 'the comparative degree of comparison', and 'the superlative degree of comparison' form the grammatical category DEGREES OF COMPARISON or simply GRADE; etc.

Just as a signifier of a morpheme is named a morph and a signifier of a lexeme a lex, a signifier that expresses a particular grammatical meaning and thus realizes a particular grammeme can be called a **gram** (or a **grammatical marker**). For example, -*ed* of e.g. *He worked* is a past tense gram; -*s* of *books* is a plural gram; -*er* of *prettier* is a comparative gram; etc.

6.2 Types of grammatical categories

Following Plungian (2000: Sec. 2.1), we will classify English grammatical categories into **syntactic** and **semantic grammemes**. The difference between them is that while syntactic grammemes produce output wordforms that differ from their input wordforms only with regard to their syntactic functioning but

Inflectional morphology

not with regard to their **referential meanings**, semantic grammemes produce output wordforms that are referentially different from their corresponding input wordforms. As an illustration of a syntactic grammeme, let us compare (87) and (88).

(87) *He met the President*
(88) *The President was met by him*

The obvious grammatical difference between the two clauses is that while (87) is in the active voice, (88) is in the passive voice. However, despite this grammatical difference, both clauses can be used to refer to one and the same meeting situation: some male person referred to in (87) as *he* and in (88) as *him* met another person (either male or female) referred to in both (87) and (88) as *the President*. The difference between the two sentences is thus not that of semantics but that of syntax: the NP *the President*, which in (87) functions as object of the predicate VP *met the President*, is in (88) **promoted** to the subject. The grammatical category VOICE is thus a syntactic grammeme that gives rise to output wordforms like *was met* of (88) that have the same referential meanings as corresponding input wordforms like *met* of (87).

As an illustration of a semantic grammeme, let us consider the referential meanings of the singular wordform *book* and the plural wordform *books*. Evidently, the two wordforms cannot refer to the same object: while the singular *book* typically refers to a single representative of the class of books, the plural *books* typically refers to more than one representative of the same class of objects. Accordingly, NUMBER is a semantic grammeme that gives rise to output nominal wordforms like *books* that have different referential meanings than corresponding input wordforms like *book*.

Semantic grammemes considerably outnumber syntactic grammemes: in English, apart from the grammatical category VOICE, only CASE can also be regarded as a syntactic grammeme. There is no referential difference between e.g. the nominative wordform *he* and the accusative wordform *him*: one and the same male person can be referred to as both *he* and *him*. By contrast, there is a referential difference between:

- an event that took place before the moment of utterance (e.g. *He met the President*) and an event that is taking place at the moment of utterance (e.g. *He is meeting the President*).

- an event that is taking place at the moment of utterance (e.g. *He is meeting the President*) and a habitual situation recurring on a regular basis (e.g. *He meets the President*).

- a real event that indeed took place (e.g. *He met the President*) and an imaginary event (e.g. *If he had met the President, he would have ...*).

- a speaker who utters a particular utterance (*I*) and the addressee of the same utterance (*you*).

- some representative of a particular class of objects (e.g. *a good book*) and the best representative of the same class (e.g. *the best book*).

Accordingly, we are justified in regarding the grammatical categories TENSE, ASPECT, MOOD, PERSON, and DEGREES OF COMPARISON as semantic grammemes, which, like NUMBER, also produce output wordforms whose referential meanings are not identical with those of their input wordforms.

6.3 Wordform-building mechanisms

These mechanisms include:

1. **inflectional affixation**
2. **analytic formation**
3. **grammatical apophony**
4. **grammatical suppletion**
5. **signifier-sharing**

6.3.1 Inflectional affixation

This wordform-building mechanism produces the majority of wordforms in English. These include:

- **nominal plural wordforms**: e.g. *books* ← *book* + *-s*
- **nominal genitive wordforms**: e.g. *father's* ← *father* + *-'s*
- **pronominal genitive wordforms**: e.g. *his* ← *he* + *-s*
- **pronominal accusative wordforms**: e.g. *him* ← *he* + *-m*
- **adjectival comparative wordforms**: *prettier* ← *pretty* + *-er*
- **adjectival superlative wordforms**: *prettiest* ← *pretty* + *-est*
- **verbal third person present tense wordforms**: e.g. *(he) reads* ← *read* + *-s*
- **verbal past tense wordforms**: e.g. *(he) worked* ← *work* + *-ed*
- **ordinal numerals**: e.g. *seventh* ← *seven* + *-th*

Inflectional morphology

In addition to these, inflectional affixation also produces **participial forms**. These include:

- so-called **participle I**, which, in combination with the auxiliary *be*, forms verbal progressive wordforms: e.g. *(he) was reading* ← *was+ read + -ing*

- so-called **participle II**, which, in combination with the auxiliaries *be* and *have*, forms passive and perfect wordforms: e.g. *(it) was created* ← *was + create + -ed* and *(he) has created* ← *has + create + -ed*

6.3.2 Analytic formation

Analytic formation is the addition of an analytic form such as, for example, the adjectival and the adverbial comparative and superlative grams *more* and *most*. For instance:

- *more beautiful / more beautifully* ← *more + beautiful / beautifully*
- *most beautiful / most beautifully* ← *most + beautiful / beautifully*

Apart from this, analytic formation can also produce wordforms together with inflectional affixation. This is true of the just mentioned:

- **progressive wordforms**: e.g. *(He) was reading* ← *was + read + -ing*
- **passive wordforms**: e.g. *(It) was created* ← *was + create + -ed*
- **perfect wordforms**: e.g. *(He) has created* ← *has + create + -ed*

6.3.3 Grammatical apophony

Similar to lexical apophony, grammatical apophony can be defined as any modification of an input lex which does not qualify as an instance of inflectional affixation. For example, the plural wordform *mice* /maɪs/ cannot be segmented into the root /ms/, expressing the meaning 'mouse', and the infix /ʌɪ/, expressing the meaning 'plurality'. Accordingly, *mice* /maɪs/ can be regarded as a product of apophony of the input singular wordform *mouse* /maʊs/. *Mice* came into existence via vowel change: /aʊ/ of *mouse* → /aɪ/ in *mice*. Similarly, the past tense wordform *met* /met/ cannot be segmented into the root /mt/, expressing the meaning 'meet', and the infix /e/, expressing the meaning 'the past tense', and hence can also be regarded as a product of apophony: /iː/ of *meet* → /e/ in *met*.

In English, grammatical apophony usually gives rise to so-called **irregular wordforms**. For example, irregular plural wordforms of nominal lexemes are those that do not take the **regular plural affix** -*s*. Apart from *mice*, these include, for example:

- *teeth* (*tooths)
- *geese* (*gooses)
- *feet* (*foots)
- *lice* (*louses)
- *men* (*mans)

Note that vowel change illustrated by *mouse* → *mice*, *tooth* → *teeth*, *goose* → *geese*, *foot* → *feet*, etc. is not the sole mechanism yielding irregular plural wordforms in English. For example, the irregular plural wordform *oxen* is a product of inflectional affixation of the input singular wordform *ox* by means of the irregular plural affix -*en*.

In addition to producing the irregular plural wordforms named above, English uses grammatical apophony for creating the following wordforms:

- **irregular third person present tense wordforms**
- **irregular past tense wordforms**
- **irregular participle II forms**

As an illustration of the first category, let us compare the present tense wordforms *have* /hæv/ and *has* /hæz/. As analyzed by Palmer et al. (2002: 1599), /hæz/ is a product of inflectional affixation of the input wordform /hæv/ by means of the regular third person suffix -*s*. However, in contrast to e.g. *save* /seɪv/, whose present tense third person wordform is /seɪvz/, the third person wordform of *have* is not */hævz/ but /hæz/. In other words, /hæz/ ← /hæv/ + /z/. That is, the affixation of the input wordform *have* by means of the regular suffix -*s* is accompanied by the reduction of the final consonant [v]. Obviously, the latter cannot be regarded as an instance of affixation (for *have* cannot be segmented into the root /hæ/ and the suffix /v/) and hence constitutes an instance of apophony. A similar example is the wordform *does* /dʌz/. Like *has*, *does* can be analyzed as a product of affixation of the input wordform *do* /du:/ by means of the regular suffix -*s*. However, in contrast to *has*, the formation of *does* is accompanied not by a consonant reduction but by a vowel change. That is, [u:] of *do* changes into [ʌ] in *does*. Again, as in the case of *has*, this modification cannot be regarded as an instance of affixation (for *do* is not segmentable into the root /d/ and the suffix /u:/) and, accordingly, constitutes an instance of apophony.

As far as past tense and participle II wordforms are concerned, we can find irregular forms that came into existence via both pure apophonies and apophonies accompanying inflectional affixation. With regard to the latter,

Inflectional morphology

consider the wordform *had* /hæd/. Similar to *has*, *had* can be analyzed as a product of affixation of *have* by means of the regular past tense suffix *-ed* accompanied by the reduction of the final consonant [v]. That is, /hæd/ = /hæv/ + /d/. Examples of pure apophonies include:

- *break* → *broke* ([eɪ] > [əʊ])
- *come* → *came* ([ʌ] > [eɪ])
- *drink* → *drank* ([ɪ] > [æ])
- *eat* → *ate* ([iː] > [e])
- *find* → *found* ([aɪ] > [aʊ])
- *give* → *gave* ([ɪ] > [eɪ])
- *hold* → *held* ([oʊ] > [e])
- *take* → *took* ([eɪ] > [ʊ])
- *wake* → *woke* ([eɪ] > [əʊ])

As in the case of irregular wordforms of nominal lexemes, irregular wordforms of verbal lexemes are not necessarily products of grammatical apophony. Consider, for example, the participle II *given* /ˈgɪv.ən/. While the past tense wordform *gave* /geɪv/ is a product of vowel change of the input present tense wordform *give* /gɪv/, *given* came into existence via affixation of *give* by means of the irregular inflectional suffix *-en*. Similarly, the irregular participle II *taken* /ˈteɪ.kən/ is not a product of apophony but of affixation of the input wordform *take* /teɪk/ by means of the irregular inflectional suffix *-en*.

Finally, observe that grammatical apophony participates in the production of the following pronominal wordforms:

- **the genitive wordform *your***
- **the genitive wordform *their***
- **the accusative wordform *them***

Your is a product of vowel change of *you*: the vowel [uː] of *you* changes into [ɔː(r)] in *your*. *Their* is a product of affixation of *they* by means of the suffix *-ir* accompanied by the reduction of the vowel [ɪ]: /ðeə(r)/ ← /ðeɪ/ + /ə(r)/. *Them* is a product of affixation of *they* by means of the suffix *-m* accompanied by the reduction of the vowel [ɪ]: /ðem/ ← /ðeɪ/ + /m/.

In contrast to the irregular nominal and verbal wordforms discussed above, these wordforms cannot be regarded as irregular pronominal wordforms: as will be shown below, the formation of genitive and accusative pronouns does not have a regular mechanism comparable to the affixation of nouns by means of *-s* and the affixation of verbs by means of *-ed*.

6.3.4 Grammatical suppletion

Grammatical suppletion is the creation of a wordform by using a signifier that has a different root than a corresponding input wordform. In English this mechanism produces:

- **pronominal genitive wordforms**: e.g. *my* ← *I*
- **pronominal accusative wordforms**: e.g. *us* ← *we*
- **pronominal person wordforms**: e.g. *you* ← *I*
- **pronominal plural wordforms**: e.g. *we* ← *I*
- **adjectival comparative wordforms**: e.g. *better* ← *good*
- **adjectival superlative wordforms**: e.g. *worst* ← *bad*
- **verbal person wordforms**: e.g. *am* ← *be*
- **past tense verbal wordforms**: e.g. *went* ← *go*
- **ordinal numerals**: e.g. *first* ← *one*

With the exception of pronominal wordforms, all other instances of grammatical suppletion are irregular wordforms. That is, the suppletive *better* and *worst* are irregular comparative and superlative adjectival wordforms (cf. *prettier* ← *pretty* + *-er* and *coldest* ← *cold* + *-est*). Similarly, the suppletive *am* is an irregular verbal present tense wordform (cf. *go* of *I go*) and the suppletive *went* is an irregular verbal past tense wordform (cf. *I worked* ← *work* + *-ed*).

As for the pronominal suppletive wordforms *my* and *us*, we cannot say that they are irregular pronominal wordforms because pronouns do not seem to have regular genitive and accusative affixes. It is true that the addition of the suffix *-s* produces the genitive wordforms *his* (← *he* + *-s*), *its* (← *it* + *-s*), *whose* (← *who* + *-s*) and the addition of the suffix *-m* produces the accusative wordforms *him* (← *he* + *-m*), *them* (← *they* + *-m*), and *whom* (← *who* + *-m*). At the same time, suppletion produces not only *my* and *us* but also *me* (← *I*), *our* (← *we*), and *her* (← *she*). In addition, as we established in 6.3.3, *your* is a product of apophony of *you*, and *their* and *them* are products of affixation of the nominative wordform *they* by means of the suffixes *-ir* and *-m* accompanied by the reduction of the vowel /ɪ/. It is not clear which of these mechanisms must be regarded as the regular genitive- and accusative-building mechanism.

As regards personal pronouns differing with regard to person, we can argue that suppletion is the default person-building mechanism. Thus the first person *I*, the second person *you*, and the third person *he*, *she*, *it* are wordforms that have different roots and, accordingly, can be analyzed as products of suppletion. Consequently, if we analyze the first person *I* as the input wordform, then both the second person *you* and the third person *he*, *she*, *it* can be analyzed as suppletive output wordforms.

Inflectional morphology

6.3.5 Signifier-sharing

Finally, signifier-sharing is the creation of a wordform by not changing an input signifier. For example, the signifier *deer* /dɪə(r)/ is both the singular and the plural wordform of the nominal lexeme DEER. Similarly, the signifier *put* /pʊt/ is both the present tense and the past tense wordform of the verbal lexeme TO PUT. In addition to nominal plural wordforms and verbal past tense wordforms, signifier-sharing also produces:

- **nominal plural genitive wordforms**: e.g. *fathers'* (← *fathers*)
- **the pronominal accusative wordform *you***: cf. *I gave you* ... and *You are* ...
- **the pronominal genitive wordform *her***: cf. *Her book* and *I gave her* ...
- **verbal imperative wordforms**: cf. *Read it!* and *I read it*
- **verbal subjunctive wordforms**: cf. *It's time that he read it* and *He read it*

Apart from these wordforms, signifier-sharing is characteristic of verbal wordforms differing with regard to the categories PERSON and NUMBER. That is, for example, we say *he goes, she goes,* and *it goes,* but *I go, you go,* and *they go.*

Consider also the signifier-sharing between past tense wordforms like *worked* and *found* of *he worked* and *he found* and participles II like *worked* and *found* of *he has worked and he has found.* Notice that this instance of signifier-sharing does not qualify as an instance of wordform-building signifier-sharing. What we call participle II is a form of a verbal lexeme like TO WORK and TO FIND that is used for producing perfect and passive wordforms. That is, a perfect wordform of TO WORK is *has worked* (i.e. a combination of the auxiliary *have* and the participle II *worked*) and a passive wordform of TO FIND is *was found* (i.e. a combination of the auxiliary *be* and the participle II *found*). In other words, the past tense wordform *worked* is not identical with the perfect wordform *has worked* and the past tense wordform *found* is not identical with the passive wordform *was found.*

6.3.6 Allowordforms

Some lexemes have more than one wordform for expressing the same grammatical meaning. In this case, we are dealing with **allowordforms**. For example, the affixed form *noisier* and the analytic formation *more noisy* are the comparative allowordforms of the adjectival lexeme NOISY. (The majority of allowordforms can be found among adjectival comparative and superlative mono- and polysyllabic wordforms, which can often be formed both inflectionally and analytically.) Similarly, *formulae* and *formulas* are the two plural allowordforms of the nominal lexeme FORMULA: the formal difference between these allowordforms is that while *formulas* is a product of affixation of the input wordform *formula* by means of the regular plural suffix *-s*, *formulae* is an instance of apophony of *formula*: the final [ə] of /'fɔːmjʊlə/ changes into [iː] in /'fɔːmjʊliː/. Finally, *webcast* and *webcasted* can be regarded as the two past tense allowordforms of the verbal lexeme TO WEBCAST 'to broadcast live over the Internet; to make viewable in real time by users of a web site' (OED / 1995): the difference between them is that while *webcasted* is an instance of affixation of the input wordform *webcast* by means of the regular past tense suffix *-ed*, *webcast* is a product of signifier-sharing.

6.3.7 Productivity

As in the case of lexeme-building mechanisms, a question can be raised as to which of the above mentioned wordform-building mechanisms are more / less productive in Present-day English. To answer this question, let us again consider the distinction between regular and irregular wordforms, which was introduced in 6.3.3. We said that regular nominal plural wordforms are those that contain the regular plural suffix *-s* and regular verbal past tense forms are those that contain the regular past tense suffix *-ed*. But why is it actually the case? How do we know that *-s* is indeed the regular plural suffix for nominal lexemes and *-ed* is the regular past tense suffix for verbal lexemes?

The answer to this question seems to be the fact that *-s* and *-ed* are the suffixes that a speaker of English is most likely to use for forming plural and past tense wordforms of recently coined nominal and verbal lexemes. That is, for example, he or she will most likely say *fake-ations, pumpkineers, (he) wifed, (she) Thomased*, etc. It is extremely unlikely that apophony, suppletion, or signifier-sharing will be used instead. The distinction between regular and irregular wordform-building mechanisms is thus a distinction between very productive mechanisms like *-s* plural affixation of nouns and *-ed* past tense affixation of verbs and rather unproductive mechanisms like apophony, suppletion, and signifier-sharing.

To study the productivity of various wordform-building mechanisms, go to the OED Advanced Search page http://www.oed.com/advancedsearch, enter '1990-' to 'Date of entry', and choose e.g. the option 'Verb'. The OED will then yield all verbal signifiers which appeared in the English language after 1990. Then carefully read each entry. If the signifier under analysis has irregular wordforms, this information will be provided in the Inflections-section of the entry. For example, the Inflections-section of the above mentioned verb TO WEBCAST contains the following information: 'Past tense *webcast, webcasted*'. This means that this lexeme has two allowordforms: the regular *webcasted* and the irregular *webcast*, which is formally identical with the corresponding present tense wordform *webcast*. Given the existence of the latter, we are justified in claiming that signifier-sharing is (at least sometimes) still used for creating verbal past tense wordforms. If you are interested in the productivity of mechanisms producing nominal wordforms, search the OED for all nominal signifiers which appeared in English after 1990 and establish (in the same way as in the case of verbs) whether at least one of them has an irregular plural form.

6.4 Syntactic grammemes in English

In this section we will discuss the syntactic grammemes VOICE and CASE. Our focus will be on the most important theoretical questions pertaining to these categories. For example, why does English need the passive voice if passive and active wordforms express the same referential meaning? Does English have the middle voice? Are *get*-passives allowordforms of *be*-passives? How many cases do we have in English and what functions do case wordforms perform in the language?

6.4.1 Why do we need the passive voice?

The answer to this question, which arises because of the non-semantic nature of the voice grammeme, is that different elements of a clause introduce different kinds of information. Among other things, there is the distinction between the **topic** and the **comment** of a clause. The topic is the center of attention of the clause, what it is about (Finegan 2004: 264). By contrast, the comment is that component of the same clause that provides some information about the topic. For example, while the active clause *He met the President* is about him who met the President, the corresponding passive clause *The President was met by him* is about the President who was met by him. In other words, in the active clause *he* is the topic and *met the President* is the comment, whereas in the passive clause *the President* is the topic and *was met by him* is the comment.

Following Mel'čuk (2006: Ch. 3), we can say that passivization is a means of changing the **communicative rank** of different pieces of information (usually, those denoting two different participants of the same event). By syntactically **promoting** the object NP *the President* to the subject of the passive clause *The President was met by him*, we also promote its communicative rank: the NP *the President*, which in the active clause is only a part of the comment *met the President*, becomes the topic of the passive clause *The President was met by him*. Similarly, by syntactically **demoting** the subject pronoun *he* to the complement of the preposition *by* of the PP *by him*, we also demote its communicative rank: *he*, the topic of the active clause *He met the President*, becomes a part of the comment *was met by him*.

6.4.2 Middle voice in English?

Middle voice is a voice that is intermediate between at least two different voices. Consider, for example, the clause *The book sells for $ 19.95*, which we already discussed in 4.3.7. In this clause the subject position is filled by the NP *The book*, which denotes the object of selling. That is, it is not the book that sells for $ 19.95. It is some people (bookstores, etc.) who sell it for $ 19.95. Given this fact, we can argue that the clause under analysis is a product of both syntactic and communicative promotion of the object NP *the book*. In this respect, the clause *The book sells for $ 19.95* does not seem to be different from the passive clause *The President was met by him*. However, in sharp contrast to the latter, the former involves the **suppression** of the performer of selling. That is, we cannot add a *by*-phrase specifying the person (the bookstore, the organization, etc.) who sells the book for $ 19.95; e.g. (89).

(89)   *The book sells for $ 19.95 by Amazon*

Given the ungrammaticality of (89), we can conjecture that *The book sells for $ 19.95* is an instance of the middle voice. The middle voice in English seems to be intermediate between the passive voice and the suppressive voice[16]. Like the former, it promotes the objects of input active clauses to subjects of output middle clauses. And like the latter, it suppresses the subjects of input active clauses.

However, at least two facts do not support the inclusion of the middle voice into the voice grammeme in English. First of all, from a formal point of view, middle clauses like *The book sells for $ 19.95* do not seem to differ from active clauses like *Amazon sells the book for $ 19.95*. In contrast to passive wordforms, which are formed with the help of the suffix *-ed* and the analytic form *be*, middle wordforms do not have a distinct grammatical marker of their own. This is the reason why many authors analyze middle clauses as clauses intermediate

---

[16] Consider, for example, the Russian clause *Собака кусается* / *Sobaka kusaetsya* 'The dog bites'. Whereas the English translation is an instance of the **absolute transitive use** – i.e. there is an understood but unexpressed object *people*, which can be added to the clause: *The dog bites people* – the Russian clause is in the **partial suppressive voice** (Mel'čuk: 205-206). The understood object *people* cannot be added to the clause: **Собака кусается людей* / *Sobaka kusaetsya lydej* / *The dog bites people*. The addition of the object *people* is suppressed by the suffix *ся* / *sya*: we can only say *Собака кусает людей* / *Sobaka kusaet lydej*, i.e. use *kusaet* 'bites' without *-sya*.

between active and passive clauses rather than between passive and suppressive clauses.

Secondly, as Huddleston (2002b: 308) points out, middle clauses are usually "concerned with whether and how [...] the subject-referent undergoes the process expressed in the verb". Because of this semantic property, middle clauses are usually 1) negative clauses like (90), 2) clauses containing modal verbs (mainly *will*) like (91), and 3) clauses containing adverbs (mainly *well* and *easily*) like (92):

(90) *The homes did not sell* (COCA)
(91) *You put her name on anything and it will sell* (COCA)
(92) *If the map sells well, it therefore immediately goes out of print* (COCA)

Finally, as Huddleston (2002b: 308) adds, a middle clause "expresses a general state, not a particular event". Consider, for example, (93) and (94).

(93) *Martinez doesn't intimidate easily* (COCA)
(94) *With a full side zip, the bag ventilates well on balmy nights* (COCA)

Clause (93) does not refer to a particular event in which somebody or something did not manage to intimidate Martinez but to a general characteristic of the person called *Martinez*. Similarly, (94) does not refer to a particular event involving somebody ventilating the bag on balmy nights but to a general characteristic of the bag.

Given these restrictions, it is clear that a number of active clauses cannot be changed into middle clauses. For example, we cannot change the active clause *He met the President* into the middle clause **The President met* meaning 'he met the President'. Likewise, we cannot change the active clause *They played football* into the middle clause **Football played* meaning 'they played football'. Both *He met the President* and *They played football* denote particular events, not general characteristics.

To conclude: the so-called middle voice exemplified by clauses like *The book sells for $ 19.95* is not a member of the set of the mutually exclusive meanings 'active voice' / 'passive voice' / 'middle voice'. The latter is not an instance of an obligatory grammatical category but a product of a particular lex-forming mechanism: lex-forming syntactics' change, which we discussed in 4.3.7. That is, some verbal lexemes that were originally realized by transitive lexes only with the course of time have acquired intransitive allolexes. For example, the transitive *sell* of clauses like *Amazon sells the book for $ 19.95* gave rise to the intransitive *sell*, which can now be used in middle clauses like *The book sells for $ 19.95*. Similarly, the transitive lex *intimidate* of clauses like *He intimidated her* gave rise to the intransitive *intimidate* of middle clauses like

Martinez does not intimidate easily. The English middle voice is a lexical, not a grammatical phenomenon.

6.4.3 What can be passivized?

The answer to this question was already (implicitly) given in 1.2.4. Passivizable are transitive clauses like *He met the President*, in which the predicative position is filled by an object: *He met the President* → *The President was met by him.*
    Transitive clauses fall into **monotransitive** and **ditransitive clauses**. The former are clauses like *He met the President* that contain only one object (the NP *the President*). The latter are two-object clauses. Consider, for example, (95).

(95)   *I don't think that people gave the President the right information* (COCA)

The clause *people gave the President the right information* is a ditransitive clause because it contains two objects: the **direct object** *the right information* and the **indirect object** *the President*. Direct objects differ from indirect objects with regard to their position: the latter can only precede the former. Consider, for instance, (96).

(96)   **People gave the right information the President*

This clause is ungrammatical because the direct object *the right information* is placed before the indirect object *the President*.
    Finally, consider the PP *to the President* of (97).

(97)   *People gave to the President the right information*

This PP expresses the same meaning as the indirect object NP *the President* of (95). Nevertheless, *to the President* is not indirect object but complement. We are justified in arriving at this conclusion because in contrast to indirect objects, the PP *to the President* can be placed either before or after the direct object *the right information*. Thus (98) is as grammatical as (97).

(98)   *People gave the right information to the President*

Another justification for the complement analysis of the PP *to the President* is that it cannot function as subject of an associated passive clause. That is, we can say *The President was given the right information (by people)* but not **To the

Inflectional morphology

President was given the right information (by people). The latter variant is ungrammatical because in this passive clause the subject position is filled by the complement phrase *to the President*, which, in contrast to the object phrase *the President*, cannot fill this position.

6.4.4 *Get*-passives

In English a 'passive' meaning can also be expressed by a combination of the verb *get* and a participle II. Consider, for example, the underlined combination in (99).

(99) *What do people in this city think of George Bush? Here's somebody who didn't campaign here. He got shellacked by Al Gore in this state* (COCA)

A clause like *He got shellacked by Al Gore in this state* is similar to the *be*-passives discussed above in that in this clause the subject position is filled by the object of shellacking (*he / George Bush*), while the performer of this action (*Al Gore*) functions only as complement in the PP *by Al Gore*. Given this similarity, a question arises as to whether the combination *got shellacked* can be regarded as an allowordform of *be shellacked*.

To answer this question, we need to establish whether *get*-passives are as obligatory as *be*-passives and whether clauses containing *get*-passives are semantically identical with corresponding clauses containing *be*-passives. In other words, does it make a difference whether George Bush *got* or *was* shellacked by Al Gore?

As far as the first question is concerned, consider (100), (101), and (102).

(100)   *His colleagues loved him*
(101)   *He was loved by his colleagues*
(102)   **He got loved by his colleagues*

The grammaticality of (101) and the ungrammaticality of (102) illustrate that *get*-passives are possible only with verbs denoting **dynamic events** (e.g. shellacking), but not with **stative verbs** like *love* (Ward et al. 2002: 1442). *Be*-passives, by contrast, are possible with both dynamic and stative verbs. That is, we can say both *He was shellacked* and *He was loved*. Accordingly, *get*-passives are less obligatory than *be*-passives.

As for the semantics of *get*-passives, consider (103), (104), (105), and (106).

(103)   *I opened the door* (neutral situation)

(104) The door <u>was opened by me</u>
(105) *The door <u>got opened by me</u>
(106) im totally shocked by how quick [...] my house <u>door got opened by the guy</u>. He must have a lot of experience cause he made it seem like nothing (http://tinyurl.com/4yww57t, a review of an emergency locksmith service)

Opening a door is a dynamic event and hence must be capable of undergoing passivization by means of *get*. However, (103) does not seem to be passivizable as (105) unless the event of my opening the door was an unusual event. Cf. (106), where the speaker is surprised by how quickly *the guy* managed to open his or her house door.

As has been pointed out in a number of studies, 'unusual' contexts favoring the use of *get*-passives are those 1) where the subject is felt to be responsible for what happened to him and 2) where the subject either benefited or suffered from what happened to him. For example, (99) is an instance of *get*-passive because, on the one hand, the subject *he* is felt to have suffered from what happened to him in this state and, on the other hand, because he himself is guilty of that: he did not campaign in this state and therefore was shellacked by Al Gore. As for (106), the speaker (and indirectly his or her house door) has also benefited from the fact that *the guy* has opened it so quickly: the speaker could quickly get into his or her house.

In summary, it appears that *get*-passives are quasi-idioms in relation to corresponding *be*-passives. The former often contain additional meanings that are not inherent in the latter. That is, for example, (99) means 'he was shellacked in this state by Al Gore and he himself is guilty of that', whereas *He was shellacked by Al Gore in this state* means only 'he was shellacked in this state by Al Gore (i.e. it may be the case that he is not to blame for this)'. Accordingly, given the semantic non-identity between clauses containing *get*- and *be*-passives and given the non-obligatoriness of *get*-passives, we can conclude this part of the section with the claim that *got shellacked* is not an allowordform of *be shellacked*. Like middle clauses, *get*-passives are a lexical, not a grammatical phenomenon.

6.4.5 Types of cases

Instances of the case grammeme such as, for example, the nominative case, the genitive case, the accusative case, etc. are sometimes classified into **straight** and **oblique cases**. The former includes only the nominative case: a nominative wordform like *I* is metaphorically the straight wordform because it is regarded as the basic or standard form of a noun or a pronoun. By contrast, all other case wordforms (e.g. the genitive *my* and the accusative *me*) have usually been metaphorized as forms that 'fall away' from their standard nominative forms (Blake 1994: 19). Accordingly, cases like the genitive and the accusative are oblique cases.

Taking this into account, we can regard the 'straight' nominative wordform (e.g. *I*) as an input wordform and all oblique forms (e.g. the accusative *me* and the genitive *my*) as output wordforms.

6.4.6 Cases in English

English personal pronouns *I, he, we, they* as well as the interrogative pronoun *who* have distinct nominative, genitive, and accusative wordforms:

- *I, my, me*
- *he, his, him*
- *we, our, us*
- *they, their, them*
- *who, whose, whom*

You and *it* have identical nominative and accusative wordforms but distinct genitive wordforms:

- *you, your, you*
- *it, its, it*

Finally, *she* has identical genitive and accusative wordforms but a distinct nominative wordform: *she, her, her*.

In contrast to pronouns, nouns have only distinct nominative and genitive wordforms: e.g. *teacher* and *teacher's*. But there does not exist a formally marked contrast between the nominative and the accusative case. For example, as illustrated by (107) and (108), one and the same wordform *teacher* can fill both the subject and the object position.

(107) The <u>teacher</u> *is able to present from the front, and is better positioned to observe pupils' response* (COCA)
(108) *I visited a social studies class and gave the* <u>teacher</u> *a map of the United States when the period was over* (COCA)

Wordforms like *teacher* of (107) and (108) are sometimes said to be in the **plain** or the **common case**, i.e. the case that combines both nominative and accusative properties.

Unlike Present-day German and Old English, Present-day English does not have the **dative case**. As stated above, the only distinctions that exist are that between the nominative, the genitive, and the accusative case (for personal pronouns and the interrogative pronoun *who*) and that between the plain case and the genitive case (for nouns). In spite of this fact, some authors often use the term **'dative shift'** describing the alternation between ditransitive clauses like *People gave the President the right information* and semantically identical monotransitive clauses containing *to*-complements like *People gave to the President the right information* (or *People gave the right information to the President*). This unfortunate term may create the wrong impression that English has the **prepositional dative case** marked by the preposition *to*. In the following we will show that this analysis is not correct.

To begin with, observe that clauses like *People gave to the President the right information* are regarded as instances of the dative shift because in languages that have distinct dative forms, the dative case marks the indirect object function. For example, we can argue that in the German clause *Man hat dem Präsidenten die richtige Information gegeben* 'people gave the President the right information', the NP *dem Präsidenten* functions as indirect object not because it precedes another object *die richtige Information* but because *dem Präsidenten* is in the dative case. Similarly, in the sentence *Ich habe ihm geholfen* 'I helped him', the pronoun *ihm* 'him' functions as indirect object because it is in the dative case. However, as far as the English language is concerned, the only criterion for distinguishing between direct and indirect objects is the position of the two objects in relation to each other. When a clause contains two objects, the indirect object precedes the direct object: thus we can only say *People gave the President the right information*, but not **People gave the right information the President*. (Objects that occur in monotransitive clauses are always direct objects: e.g. even though *him* of the English *I helped him* is semantically identical with the German indirect object *ihm* of *Ich habe ihm geholfen*, the English *him* is the direct object of the clause *I helped him*.) As for monotransitive clauses containing *to*-phrases, we already pointed out that PPs headed by the preposition *to* such as *to the President* are not indirect objects but complements. On the one hand, this is so because they can occur both before and after direct objects: we can say both *People gave to the President the right information* and *People gave the right information to the President*. On the other hand, this is so because *to*-phrases cannot serve as subjects of associated passive clauses: we cannot say **To the president was given the right information*. Accordingly, since *to*-phrases cannot function as indirect objects (which in languages that have the dative case are expressed by NPs in the dative), the preposition *to* cannot be a dative gram. Hence English does not have the dative case.

6.4.7 Functions of case

Since case is a syntactic grammeme, it is not surprising that wordforms differing with regard to case often mark different syntactic functions. For example, as was already pointed out in 1.2.4, the subject position can be filled by a pronoun in the nominative (e.g. *He is a good guy*), but not by a pronoun in the accusative (e.g. **Him is a good guy*). In contrast, the object position can be filled by a pronoun in the accusative (e.g. *I helped him*), but not by a pronoun in the nominative (e.g. **I helped he*).

With regard to genitive wordforms, consider (109), (110), (111), (112), (113), (114).

(109) <u>Mike's</u> *eyes widened and he asked Leo, 'Isn't the ground starting to tilt?* (COCA)
(110) <u>His</u> *being out of control was her fault* (COCA)
(111) *Our favorite find? Frozen biscuits. They taste nearly as good as* <u>grandma's</u>, *minus all the work* (COCA)
(112) *He was a friend of* <u>father's</u> (COCA)
(113) *Much as they like the 'sameness' of routine to end the day, kids will often happily accommodate two rituals, especially if one is* <u>mom's</u> *and one is* <u>dad's</u> (COCA)
(114) *It was a* <u>girl's</u> *voice, behind him: strong and bright and sure* (COCA)

These examples illustrate what Payne and Huddleston (2002: 467) call **six types of genitive constructions** in English. The genitive wordform *Mike's* of (109) functions as determiner of the head noun *eyes*. Mike's eyes are particular representatives of the class of objects called *eyes*: those that belong to Mike.

The genitive pronoun *his* of (110) functions as subject of the embedded **non-finite clause** *His being out of control* (which in turn functions as subject of the larger clause *His being out of control was her fault*, into which it is embedded): *His being out of control* means 'he was out of control'.

The genitive wordform *grandma's* of (111) is a quasi-idiom whose signified contains not only the signifieds 'grandmother' and 'the genitive case' but also the idiomatic meaning 'biscuits' recoverable from the preceding clause *Frozen biscuits*. Precisely because of its quasi-idiomatic signified, *grandma's* of (111) functions as a **fused determiner-head phrase**, i.e. a phrase in which both the determiner and the head are expressed by one and the same element.

The genitive wordform *father's* of (112) functions as **post-head modifier** of the head noun *friend*. As the term makes clear, a post-head modifier is a modifier occurring after the head of a phrase, not before it.

The genitive wordforms *mom's* and *dad's* of (113) function as complements of the auxiliary verb *is*.

Finally, *girl's* of (114) exemplifies a genitive construction that is often called **classifying** or **descriptive genitive**. In contrast to *Mike's* of *Mike's eyes*, *girl's* of *a girl's voice* does not identify the voice as the voice of a particular girl but describes it as a voice that is generally characteristic of a young woman. Classifying genitives like *girl's* of *a girl's voice* function as **pre-head modifiers** of head nouns in NPs.

As we established in 1.2.5, sometimes one and the same syntactic position can be filled by more than one case wordform. For example, we can say both *It is I who did it* and *It is me who did it, It is he who did it* and *It is him who did it*, etc. As we said, nominative and accusative pronouns filling the complement position in sentences like *It is I / me who...* differ with regard to their sociolinguistics. The nominative wordforms make these sentences sound more formal than the corresponding accusative wordforms. Given this difference, we can say that case in English has one more important function: the **sociolinguistic** (or the **stylistic**) **function**.

Indeed, the complement position in sentences like *It is I / me who...* is not the only syntactic position where a speaker of English can use more than one case wordform. Consider, for example, the variation in the object position between the nominative *who* of (115) and the accusative *whom* of (116).

(115) *Who did you vote for?* (COCA)
(116) *Whom did you vote for [...]?* (COCA)

In this position, it is the accusative *whom* that makes a clause like (116) sound more formal than a clause like (115), in which the object position is filled by the nominative *who*.

6.4.8 Semantic function of case?

Given what we have said about the syntactic nature of the case grammeme, this title may seem a **contradictio in adjecto**: CASE cannot have a semantic function because it is a syntactic grammeme. Nevertheless, Quirk et al. (1985: 321-322) speak of **'genitive meanings'** which can be expressed by genitive wordforms in English. For example, the meaning 'possession' inherent in *Mike's house*: 'Mike's house' = 'the house that Mike possesses'. The possessive meaning is often seen as the default genitive meaning, so that the English genitive case is sometimes referred to as the **possessive case**.

Apart from this meaning, the English genitive is said to express **subjective** and **objective meanings**. For instance, the signified of the NP *Mike's decision* can be represented as the clause *Mike made a decision*, where *Mike* is the subject. By contrast, the signified of the NP *Mike's arrest* can be represented as

Inflectional morphology

the clause *The police arrested Mike*, where Mike is the object. (Instead of using the syntactic terms 'subject' and 'object', it is perhaps better to describe these meanings in terms of **semantic roles**: Mike is the **agent** of making a decision and Mike is the **patient** of being arrested by the police.) Also, the genitive case in English can express:

- **partitive meanings**: e.g. *Mike's eyes* = 'eyes that are part of Mike'
- **temporal meanings**: e.g. *New Year's Eve* = 'evening before the New Year'
- **measure meanings**: e.g. *an hour's discussion* = 'a discussion that lasted one hour'

This book argues against the treatment of these meanings as genitive meanings. These meanings are not inherent in the genitive gram *-'s* itself but in NPs like *Mike's house*, *Mike's decision*, *Mike's arrest*, *Mike's eyes*, *New Year's Eve*, and *an hour's discussion*. That is, all these NPs are quasi-idioms whose signifieds contain additional idiomatic meanings that are not inherent in either of their components' signifieds. For example, the NP *Mike's house* is a quasi-idiom whose signified contains not only the signifieds 'Mike' and 'house' but also the idiomatic meaning 'possession', which is not inherent in either *Mike* or *house* or the genitive gram *-'s*. Similarly, the NP *Mike's decision* is a quasi-idiom whose signified does not only contain the signifieds 'Mike' and 'decision' but also the idiomatic meaning 'made by', which is not inherent in either *Mike* or *decision* or the genitive gram *-'s*.

The existence of multiple 'genitive meanings' corroborates the quasi-idiomatic analysis of all these NPs. If there were such thing as an obligatory genitive meaning, then all genitive wordforms would carry that meaning in all (or at least the majority of the) contexts where they occur. However, as the NPs under consideration demonstrate, phrases containing genitive wordforms can express more than one genitive meaning: 'possession', 'agent', 'patient', 'being a part of', 'temporal location', 'measure', etc. Each of these meanings is not inherent in the genitive suffix *-'s* but is a product of quasi-idiomatization of NPs containing genitive wordforms.

To conclude: like the nominative and the accusative, the English genitive case is a syntactic case, which does not carry any (genitive) meaning of its own. Accordingly, we can reiterate the characterization of CASE as a syntactic grammeme that produces output wordforms like *Mike's* that have the same referential meaning as corresponding input wordforms like *Mike*. (Since this book is concerned with English morphology, we can ignore languages that have **semantic cases**. For example, in Finnish there is the **inessive case**. This case does not only serve to mark a particular syntactic function of a noun but also to express the meaning 'location within the referent' (Lyons 1968: 299). For example, the noun *talo-ssa*, where *-ssa* is an inessive gram, means 'inside the house'. English lacks semantic cases, so that we are fully justified in regarding the case grammeme in English as a syntactic grammeme).

6.5 Semantic grammemes in English

Having discussed the syntactic grammemes VOICE and CASE, we can proceed to the semantic grammemes TENSE, ASPECT, MOOD, PERSON, NUMBER, DEGREES OF COMPARISON, and NUMERICAL QUALIFICATION.

6.5.1 Typology of temporal meanings

Tense is the grammatical category that is concerned with the relation holding between an event expressed by a predicate VP and its **tense locus**, i.e. the reference point in relation to which the event is located in time (Frawley 1992: 340). Tenses that are concerned with the relation holding between events and the **moment of utterance** (i.e. the primary tense locus) are **simple tenses**. These include the **past simple tense** and the **present simple tense**.

If the event under analysis precedes the moment of utterance, it is in the past tense. If the event coincides with the moment of utterance, it is in the present tense. As an illustration of this difference, compare the temporal meanings of *ask* in (117) and (118).

(117)   *Between the beer course and the pizza, I <u>asked</u> you to describe the perfect woman* (COCA)
(118)   *So I <u>ask</u> you to keep quiet about this thing [...]* (COCA)

I asked you to describe the perfect woman of (117) is in the past tense: the event described by the clause took place before the speaker uttered (117), i.e. he or she first said *Please describe the perfect woman* and then by uttering (117) reminded the hearer(s) about this request. By contrast, *I ask you to keep quiet about this thing* of (118) is in the present tense: the event described by the clause coincides with the moment of utterance, i.e. at the moment when the speaker uttered (118), he or she asked the hearer(s) to keep quiet about some thing.

Tenses which have more than one tense locus are **perfect tenses**. Compare, for example, (119) and (120).

(119)   *I saw the movie*
(120)   *I had seen the movie*

Both (119) and (120) denote events which were accomplished before the moment of utterance. That is, the speaker first saw the movie and then said *I saw the movie* / *I had seen the movie*. However, while (119) is located in the past only in relation to the moment of utterance, (120) also expresses anteriority in relation to an additional tense locus: either some other event or some specified time in the past; e.g. (121) and (122).

(121)   *I had seen the movie before my mother came*
(122)   *I had seen the movie by 10 o'clock yesterday*

The **past perfect tense**, exemplified by (121) and (122), can be contrasted with the **present perfect tense**. Consider, for example, (123).

(123)   *I have seen the movie*

Like (119) and (120), this clause denotes an event that took place before the moment of utterance: the speaker first saw the movie and then said *I have seen the movie*. The difference between (123) and both (119) and (120) is that the past event expressed by (123) is somehow relevant in the present. Consider, for example, (124).

(124)   *I have seen the movie, and can confirm that it might be the most staggering filmic achievement of this or any other century* (http://tinyurl.com/66n68ja)

Here the use of the present perfect tense can be explained in the following way. It is relevant in the present that the speaker saw the movie in the past because owing to this, he or she now knows which movie can be considered the most staggering filmic achievement of this or any other century. The difference between the past perfect tense and the present perfect tense is thus that while the secondary tense locus of the past perfect tense is in the past, the secondary tense locus of the present perfect tense is in the present. (Hence the term 'present perfect', even though this tense is also associated with past events.)

In addition to the simple tense–perfect tense distinction, there is also the distinction between **absolute** and **relative tenses**. Absolute tenses are tenses whose tense locus is the moment of utterance. Both simple tenses and perfect tenses qualify as absolute tenses because both the former and the latter characterize the relation between an event and the moment of utterance. The only difference is that in the case of simple tenses, the moment of utterance is the sole tense locus relative to which the event is located in time, whereas perfect tenses have two tense loci: the moment of utterance and some other event / some specified time. By contrast, relative tenses do not depend on the moment of utterance as their tense locus. Consider, for example, the temporal meanings inherent in the underlined non-finite clauses in (125), (126), (127), (128).

(125)   I'm glad <u>to have the opportunity to apologize to you, Ms. Shipman, in person</u> (COCA)
(126)   I'm glad <u>to have had the opportunity to meet you, Ranger Strong [...]</u> (COCA)
(127)   I enjoyed <u>writing songs</u> (COCA)
(128)   Many other playwrights have written better, but no one has enjoyed <u>having written more</u> (COCA)

Non-finite clauses are clauses that contain **non-finite forms** such e.g. the **infinitive** *have* of (125) and the **gerund** *writing* of (127). Among other things, these are different from **finite forms** (e.g. *enjoyed* of (127)) with regard to their temporal meanings. The tenses of non-finite clauses are always relative tenses whose tense locus is not the moment of utterance but the event expressed by the finite clause. For example, the infinitival clause in (125) is an instance of the **indefinite infinitive**, i.e. a relative tense expressing **simultaneousness**. That is, the event denoted by the non-finite clause *to have the opportunity to apologize to you, Ms. Shipman, in person* is simultaneous with the event denoted by the finite clause *I'm glad*. In other words, the speaker of (125) is glad that he or she at the moment of utterance has the opportunity to apologize to Ms. Shipman. The same can be said about the **indefinite gerund** exemplified by the gerundial clause in (127). The event denoted by the non-finite clause *writing songs* was simultaneous with the event denoted by the finite clause *I enjoyed*.

Indefinite infinitives and indefinite gerunds must be distinguished from **perfect infinitives** and **perfect gerunds**. For example, the infinitival clause in (126) is a perfect infinitive which expresses that the event denoted by the non-finite clause *to have had the opportunity to meet you, Ranger Strong* is anterior to the event denoted by the finite clause *I'm glad*. Likewise, the gerundial clause in (128) is a perfect gerund which expresses that the event denoted by the non-finite clause *having written more* is anterior to the event denoted by the finite clause *no one has enjoyed*.

6.5.2 No future tense in English

It is often argued that the English language lacks the future tense. The reasons for this provided by the authors who share this view can be summarized in the following way:

- No inflectional affixes are used for forming future tense wordforms. Cf. e.g. *(he) worked* and *(he) will work*.

- There is an overlap between the present and the future tense, i.e. one and the same wordform can be used for expressing both present and future meanings.

- The meaning 'the future tense' is a **modal** rather than a temporal meaning.

As for the absence of inflectional grams, it is indeed the case that in contrast to e.g. Lithuanian, which has separate inflectional encodings for the past tense, the present tense, and the future tense (Frawley 1992: 359), English has only the past tense regular affix *-ed* and the present tense (third person singular) suffix *-s*. However, it is doubtful that this fact alone suffices to conclude that English lacks the future tense: as we have learned in 6.3, wordform-building in English relies not only on inflectional affixation but also on analytic formation (e.g. *more beautiful*). Accordingly, a future tense form like *will work* can be regarded as an analytic wordform consisting of the input present tense wordform *work* and the analytic auxiliary verb *will*.

With regard to the overlap between the present tense and the future tense, consider (129) and (130).

(129)   *The next solar eclipse is on July 11 2011* (http://tinyurl.com/45ytebe)
(130)   *Hockey World Cup: England meets Germany in semis tomorrow* (http://tinyurl.com/3bbs6tm)

These examples serve to illustrate that present tense wordforms may have a future reference. As pointed out by Huddleston (2002b: 132), this is true of scientifically calculable cyclic events such as e.g. a solar eclipse of (129) and scheduled events such as e.g. a hockey game of (130). But again this fact is not sufficient to claim that English does not have the future tense. Thus in addition to the present–future overlap, there are also the present–past and past–future overlaps. Consider, for example, (131) and (132).

(131)   [...] <u>the Bible says</u> that having respect for God is the beginning of wisdom (COCA)
(132)   *I figured it was best to wait <u>until he was ready</u>* (COCA)

In (131), the present tense wordform *says* has a past tense reference: *the Bible says that having respect for God is the beginning of wisdom* means 'the author of the Bible said (long before the moment of utterance of (131)) that having respect for God is the beginning of wisdom'. By using the present tense wordform *says*, the speaker of (131) emphasizes the present-day importance of what the author of the Bible said long ago. In (132), the past tense wordform *was ready* refers to a future event viewed from the past, i.e. the person referred to as *he* will be ready after the speaker of (132) figured it was best to wait. Obviously, these overlaps provide no justification for the claim that the English language lacks either the present or the past tense.

Finally, let us consider the claim that the future tense meaning is not a temporal but a modal meaning. The point here is that no matter how certain you might be about your future plans, a future event is always an event that may or may not take place. Consider, for example, (133).

(133)   *I will travel to Mexico soon* (COCA)

Even if the speaker has already booked the flight to Mexico and has already made a hotel reservation, (133) remains a **prediction**: it cannot be excluded that he or she will die in a car accident while driving to the airport and thus will not be able to travel to Mexico.

This leads us to the conclusion that the future tense is different from both the past tense and the present tense in that while the latter denote **real events** that either coincide with or precede the tense locus, the former denotes **unreal events** that may or may not take place in the future. Accordingly, the future tense must be analyzed not as a member of the tense grammeme but as an instance of so-called **epistemic modality**, i.e. a type of modality which is concerned with "the degree of certainty the speaker has that what s/he is saying is true" (Haan 2006: 29). That is, the future *will* of (133) is a modal verb which expresses more certainty than e.g. the epistemic *may* of *I may travel to Mexico soon* 'it's possible that I will travel to Mexico soon. But I am not certain about this'.

6.5.3 Idiomatic uses of temporal wordforms

As we argued in 2.5.4, wordforms containing inflectional affixes such as the past tense suffix *-ed* can express idiomatic meanings. For example, the use of past tense wordforms in **politeness-related contexts** (Huddleston 2002b: 138) such as e.g. *I wanted to ask you to tell the story of your father taking you to see the Brooklyn Dodgers in 1947* meaning 'I want to ask you to tell this story now' is a case in point. Here *-ed* does not characterize the event as a past event that was valid before the moment of utterance (i.e. 'some time ago I wanted to ask you to tell this story, but at present I no longer want to hear it') but as a past event that continues to the present. That is, a person who says *I wanted to ask you* (most likely) began to experience this wish prior to the moment of utterance. However, at the moment of utterance he still wants the addressee to tell the story. Past tense wordforms like *wanted* of politeness-related contexts like *I wanted to ask you* can thus be analyzed as quasi-idioms in relation to 'normal' past tense wordforms such as e.g. *wanted* of (134) denoting past tense events which do not continue to the present:

(134)   A year ago I <u>wanted</u> to end my life. I am now 50 and looking forward to the next 50 (COCA)

Evidently, the speaker of (134) does not want to end his or her life anymore. By contrast, *wanted* of *I wanted to ask you* contains not only the meaning 'the past tense', inherent in the past tense suffix *-ed*, but also the idiomatic meaning 'continuation to the present'.

Consider also the temporal meaning of the present tense wordform *becomes* in (135).

(135)   August 3 – Bolivia <u>becomes</u> the first South American country to declare the right of indigenous people to govern themselves (http://en.wikipedia.org/wiki/2009)

(135) is taken from Wikipedia's 2009 historical chronicle. It thus describes a past event that took place in August 2009. The past tense use of the present tense wordform *becomes* of (135) can thus be analyzed as a full-idiom in relation to *becomes* of e.g. *It now becomes apparent that ...*, where it denotes an event coinciding with the moment of utterance.

Finally, consider (136).

(136)   *I have lived in New York*

This clause is ambiguous between the following two readings: 'I lived in New York in the past and this is somehow relevant in the present' and 'I moved to New York in the past and still live there'. The former is known as the **non-continuative present perfect**; the latter as the **continuative present perfect**. According to Huddleston (2002b: 141), "the non-continuative reading of the perfect is much more frequent, and can be regarded as the default one". Indeed, it is more likely that a speaker of English will interpret (136) non-continuatively, i.e. as an event that does not continue to the present. A continuative interpretation will occur only in contexts like *I have lived in New York for ten years*, where the continuative reading is reinforced by the duration adjunct *for ten years*. Taking this into account, we can analyze the continuative present perfect as an idiomatic use of the present perfect tense. The continuative perfect is a semi-idiom in relation to the non-continuative perfect. While the latter can be segmented into the features [beginning in the past], [accomplishment in the past], [present relevance], the former is segmentable into the features [beginning in the past] and [continuation in the present].

6.5.4 Typology of aspectual meanings

Aspect is the grammatical category which is concerned with aspectual meanings (sometimes referred to as **Aktionsarten**). Following Maslov (1973; 1978), Frawley (1992: Ch. 7), Plungian (2000: 292-308), and Huddleston (2002b: 118-125), we will classify aspectual meanings into those relating to:

1. **qualitative aspectuality**
2. **quantitative aspectuality**
3. **phase aspectuality**

With regard to qualitative aspectuality, which is concerned with the inherent properties of an event expressed by a predicate VP, events can be classified into **dynamic events** and **static events**. These are different from each other with regard to the following grammatical properties:

- The present simple combines with static events but not with dynamic events: i.e. e.g. we can say *He knows Mike* meaning 'He knows Mike at the moment of utterance' but not **He reads a book* meaning 'He is reading a book at the moment of utterance'.

- Only dynamic events can be progressivized: cf. e.g. *He is reading a book* and **He is knowing Mike*.

- Only dynamic events can occur in a *wh*-cleft: cf. e.g. *What he did next was read a book* and **What he did next was know Mike*.

Dynamic events can be further classified into **achievements** (e.g. *He found a book*) and **processes** (e.g. *He read a book*). These are different from each other with regard to the following properties:

- Verbs like *begin, finish, continue*, etc. combine with processes but not with achievements: cf. e.g. *He began to read a book* and **He began to find a book*.

- Only processes can be progressivized: cf. e.g. *He is reading a book* and **He is finding a book*.

- Duration adjuncts like the PP *for an hour* combine freely with processes but not with achievements: cf. e.g. *He read a book for an hour* and **He found a book for an hour*.

Finally, processes can be further classified into **telic events** (e.g. *We went to the beach*) and **atelic events** (e.g. *He teaches history at Yale University*). The differences are as follows:

- Telic events are begun in order to be finished: e.g. we go to the beach in order to be at the beach. As soon as we are at the beach, we no longer continue going to the beach. Atelic events, by contrast, do not have such terminal points that terminate the realization of an event as soon as they are achieved: e.g. one can teach history at Yale as long as one wishes to do so.

- We can say *It took us an hour to go to the beach* but not **It took him a year to teach history at Yale University*.

With regard to quantitative aspectuality, which is concerned with the question of how many times a given event was realized, events can be classified into:

1. **single or semelfactive events**: e.g. *Mike knocked (once) at the door*

2. **multiple or iterative events**: e.g. *Mike knocked at the door for a minute*

3. **permanent events that always take place**: e.g. *CNN broadcasts 24 hours a day*

Multiple events can be further classified into the following types:

a) **multiplicative events**, consisting of a set of semelfactive events which are repeated instantaneously: e.g. *Mike knocked at the door for a minute*

b) **distributive events**, applying to members of the same group one after another: e.g. *Mike knocked at several doors*

c) **habitual events**, recurring on a more or less regular basis; e.g. *He eats one meal a day*

Finally, phase aspectuality is concerned with the question of which phase of an event was realized at some specified time. According to Plungian (2000: 297), event phases or stages include:

1. **preparation / prospective phase**: e.g. *I am going to read a book*

2. **beginning / inchoative phase**: e.g. *I begin to read a book*

3. **cessation / terminal phase**: e.g. *I finish reading a book*

4. **middle phase** (i.e. the phase between the beginning and the cessation of an event): e.g. *I am reading a book*

5. **result / resultative phase**: e.g. I finish reading a book and go to the movie to see the film based on the book

The stages 'beginning', 'cessation', and 'the middle phase' constitute the **internal stages** of an event, while the preparatory and the resultative phases are the **external phases**. This is so because it can be argued that, for example, the opening of a book, which is usually a part of the preparatory phase of the reading event (we cannot start reading a book without opening it), is not really a part of the reading event itself: at the moment when we are opening a book, we are not reading it.

6.5.5 Aspects in English

English has a number of **aspectualizers** that are used for expressing the above named aspectual meanings. But only one of them – a combination of the auxiliary *be* and a participle I like *working* – can be regarded as a marker of an obligatory grammatical meaning. This textbook thus accepts the traditional view that English has only one aspect: **the progressive aspect**.

The progressive is the aspect which expresses the meaning 'the realization of the middle phase of an event'. For example, the progressive clause *He was reading a book* expresses the meaning 'the realization of the middle phase of the reading event', i.e. at the moment when the speaker's attention was attracted by this event, the subject *he* was in the middle of reading a book. He was neither beginning nor finishing reading.

In 6.5.4 we have learned that the progressive aspect does not combine with states (e.g. **He is knowing Mike*) and achievements (e.g. **He was finding a book*). With regard to the latter, the explanation is that achievements are **instantaneous events**, i.e. events that lack not only the middle phase but also all other internal phases discussed above. That is, it is impossible **to be finding a book* because it is not possible either **to begin to find a book* or **to finish finding a book*. You either find it or not. With regard to states, consider (137) and (138).

(137)   ???*He began to know Mike*
(138)   ???*He finished knowing Mike*

Both (137) and (138) are hardly possible in English because both the inchoative and the terminal phase of a static event are perceived as external stages of the event. That is, the inchoative meaning inherent in (137) and the terminal meaning inherent in (138) are usually expressed by sentences like *He got acquainted with Mike* and *He forgot Mike*, i.e. sentences which do not mention the knowing event itself. The impossibility of the progressivization of the state *He knows Mike* can thus be attributed to the fact that the knowing event consists of only one middle phase, i.e. there is no need to progressivize *He knows Mike* because, by virtue of being a state, this event already conveys the meaning 'the middle phase'.

Sometimes it is argued that English has the **habitual aspect** expressed by *used to* of e.g. (139) and the **prospective aspect** expressed by *be going to* of e.g. (140).

(139)   *I used to walk down the street in New York City [...]* (COCA)
(140)   *I was going to get there* (COCA)

Indeed, (139) expresses that the action of walking down some street in New York City was repeated by the speaker on a fairly regular basis, and (140) expresses that the speaker planned to get to some place. However, both *used to* and *be going to* are much less obligatory than the progressive marker *be* + participle I. With regard to *used to*, observe that a repetition of an action in the past can also be expressed with the help of *would*. For example, (141).

(141)   *[...] every day he would go out in the backyard and pass the tales on to whatever animals happened to be lounging about* (COCA)

And in the case of a present tense event, a habitual meaning is usually expressed with the help of the present simple tense. That is, a clause like *He reads a book* means 'He habitually repeats the action of reading a book'. By contrast, a clause like *He used to read a book* can only mean 'This series of events took place in the past (not in the present)'.

Similarly, as Comrie (1976: 64) points out, English has several prospective aspectualizers. For example, *be about to* of (142) and *be on the point of* of (143).

(142)   *I remember the day I was about to teach my first course in political psychology* (COCA)
(143)   *I was on the point of quitting for years* (COCA)

Also, it is worth mentioning that *be going to* "is characteristic of relatively informal style" (Huddleston 2002b: 211), whereas all other grams that have been discussed so far are stylistically neutral.

Given the non-obligatory character of both *used to* and *be going to*, we are justified in concluding that these idiomatic combinations are not grams of obligatory aspectual meanings but lexical aspectualizers for expressing optional habitual and prospective meanings.

Finally, let us consider the possibility of regarding the present simple non-progressive as a marker of the habitual aspect in English. This conjecture may arise because, as said above, the present simple tense exemplified by *He reads a book* expresses the habitual meaning 'He regularly reads some book' and, as we established in 6.5.4, habituality is an instance of quantitative aspectuality. To answer the question of whether the English present simple tense is indeed an aspectual rather than a temporal category, we first need to consider the question: why is it actually the case that a clause like *He reads a book* can only be associated with an aspectual habitual meaning rather than with the temporal meaning 'an event that coincides with the moment of utterance'?

The answer to this question relates to what we have learned about the stages of an event in 6.5.4. A dynamic event like reading a book has three internal stages: the beginning, the cessation, and the middle stage. It is obvious that the moment of utterance cannot coincide with all three stages of this event: it takes much more time to read a book than it does to say *He reads a book*. The moment of utterance can coincide only with one of these stages. If it coincides with the middle stage, we use the progressive wordform *is reading*. Accordingly, we can say that *He reads a book* expresses an aspectual habitual meaning because due to its dynamic aspectuality, it cannot express the temporal present tense meaning 'an event taking place at the moment of utterance'.

At the same time, note that some dynamic events do have a non-habitual interpretation when used in the present simple tense. This is true of e.g. running commentaries such as (144):

(144)   Now on the board Semyonov, number 30, back to Fetisov, to Kasatonov, <u>shoots</u>, he <u>scores</u>!!! (http://tinyurl.com/6316pls, Canada–USSR, Canada Cup 1987, Final, Game 1)

At the moment when the commentator said *shoots, he scores!!!*, Kasatonov (i.e. a Soviet ice hockey player) was indeed shooting and scoring a goal. In other words, the shooting and scoring events coincided with the moment of utterance.

Recall also the use of the present simple wordform *say* in *I ask you to keep quiet about this thing*. As stated in 6.5.1, by uttering this sentence the speaker did indeed ask the addressee(s) to keep quiet about some thing. Again, we can say that the event of asking other people coincided with the moment of utterance. Both these facts support the traditional analysis of the present simple tense as a temporal rather than an aspectual category.

6.5.6 Moods in English

Similar to tense and aspect, which can be defined as grammatical categories concerned with the expression of temporal and aspectual meanings, mood can be defined as the grammatical category concerned with the expression of **modal meanings**.

Traditionally, modality is classified into:

1. **root modality**, which is associated with the meaning 'ability': e.g. *Sarah can speak three languages* 'Sarah has the ability to speak three languages'.

2. **deontic modality**, which is associated with the meaning 'obligation': e.g. *Sarah must speak three languages* 'Sarah is required to have the ability to speak three languages'.

3. **epistemic modality**, which is associated with the meanings 'certainty' and 'possibility': e.g. *Sarah will speak three languages* 'It is possible that in the future Sarah will have the ability to speak three languages'.

As in the case of aspectual meanings, modal meanings can be expressed in a number of ways. In addition to modal verbs like the above named *can*, *must*, and *will*, modal (especially epistemic) meanings are also often expressed with the help of **modal adverbs** like *probably* and *maybe* of (145) and (146) and so-called **modal tags** like *I believe* and *I guess* of (147) and (148).

(145)   Barry is <u>probably</u> at home watching our games (COCA)
(146)   <u>Maybe</u> the danger lies within ourselves (COCA)
(147)   <u>I believe</u> I developed better negotiating skills in the business world than a lot of people who've been around here for years [...] (COCA)
(148)   <u>I guess</u> he's been busy (COCA)

However, none of these **modalizers** can be regarded as obligatory grammatical markers in English. Thus instead of saying *Sarah must speak three languages*, we can say *Sarah is required to speak three languages* or *It is necessary that Sarah speak three languages*. And instead of saying *Sarah may speak three languages* meaning 'it is possible that Sarah speaks three languages but I am not certain about it', we can say *I guess Sarah speaks three languages* or *Sarah probably speaks three languages*.

The only modal grams in English are those modalizers that characterize an event as an instance of one of the following traditionally recognized moods:

1. **the indicative mood**
2. **the imperative mood**
3. **the subjunctive mood**

The indicative mood is described by Khaimovich and Rogovskaya (1967: 142) as "a fact mood [which] serves to present an action as a fact of reality". For example, a clause like *He is reading a book* denotes an event that, according to the speaker, coincides with the moment of utterance. Similarly, a clause like *He was reading a book* denotes an event that, according to the speaker, preceded the moment of utterance.

A notable feature of the indicative mood in English is that there are no grams that serve to express indicative meanings only. Consider, for example, (149).

(149)   *He studies engineering and business management at Chongqing University [...]* (COCA)

This sentence can only be interpreted indicatively, i.e. as a sentence denoting an event that is still valid at the moment of utterance. That is, at the moment of utterance the subject *he* was still a registered student of engineering and business management at Chongqing University. What justifies this analysis is that the predicate VP is headed by the wordform *studies*, which is a product of affixation of *study* by means of the suffix *-s*. However, as we have learned in 2.1, this suffix is a portmanteau morph, which cumulatively expresses more than one grammatical meaning. That is, in addition to the meaning 'the indicative mood', it also expresses the meanings 'the present tense', 'the non-progressive aspect', 'the active voice', 'the third person', etc. The same is true of all other indicative wordforms such as e.g. *(He) studied, (He) has studied, (It) was studied*, etc. In all of them, the indicative meaning is expressed cumulatively, i.e. together with other grammatical meanings such as 'tense', 'aspect', 'voice', 'number', etc.

In contrast to the indicative mood, the imperative mood has several genuinely imperative formal characteristics. One is that the majority of imperative clauses have covert second person subjects. For example, (150) can be said to mean 'You go to the bedroom' and the signified of (151) can be analyzed as 'You replace the vacuum bag once a month'.

(150)   *Go to the bedroom* (COCA)
(151)   *Replace the vacuum bag once a month* (COCA)

Inflectional morphology

In other words, it is the addressee(s) of (150) and (151) that are expected to perform the actions of going to some bedroom and replacing the vacuum bag once a month.

The second formal characteristic is that the VPs of imperative clauses are headed by verbs in the **plain form**. (Huddleston 2002b: 83). This is the citation form of a verb (*go, replace, follow*). Usually it is identical with the present tense wordform. However, in contrast to the latter, the former does not receive the affix *-s* when used with third person subjects like *he* and *she*. Compare, for example, (152) and (153).

(152) *Everybody follows that* (COCA)
(153) *Come on. Everybody follow me* (COCA)

Clause (152) is in the indicative mood: *Everybody follows that* = 'According to the speaker, at the moment of utterance everybody followed that'. By contrast, (153) is in the imperative mood: *Everybody follow me* = 'The speaker asks the addresses to follow him'. We are justified in arriving at this conclusion because the predicator *follow* is in the plain form: it does not receive the third person suffix *-s*, which it receives in an indicative clause like (152).

Semantically, the imperative mood is associated with the meanings 'order', 'command', 'request', 'instruction' and the like. In addition to these core imperative meanings, imperative clauses can also be used for expressing wishes and conditions. Consider, for example, (154) and (155).

(154) *Have a happy Thanksgiving* (COCA)
(155) *In spite of repeated warnings you insist on posting in capitals! <u>Do this again</u> and you will receive a formal warning*
       (http://tinyurl.com/5uz4pee)

The imperative clause (154) means 'I wish you a happy Thanksgiving' and the imperative clause in (155) means 'If you do this again, you will receive a formal warning'.

Finally, let us turn our attention to the subjunctive mood. If the indicative is a fact mood, the subjunctive is a non-fact mood, i.e. a mood that represents an event as "something imaginary, desirable, problematic, contrary to reality" (Khaimovich and Rogovskaya 1967: 150-151). The subjunctive mood is often classified into **subjunctive I** and **subjunctive II**. (Subjunctive I is sometimes called **present subjunctive** and subjunctive II **past subjunctive**. (Quirk et al. 1985: 155). Huddleston (2002: 88) uses the term 'subjunctive mood' only in connection with subjunctive I, whereas subjunctive II is referred to by him as the **irrealis mood**.) Compare, for example, (156) and (157).

(156)   I would suggest that <u>he start</u> helping the people of the United States that [have] worked and paid in taxes all their lives (COCA)
(157)   I wish <u>you were</u> here (COCA)

The clause *he start helping the people of the United States [...]* of (156) is an instance of subjunctive I, i.e. a mood which represents events as problematic but not as entirely contradicting reality (Ganshina and Vasilevskaya 1964: 203). Thus it is possible that the subject *he* will indeed start helping the people of the United States. By contrast, the clause *you were here* of (157) is an instance of subjunctive II, i.e. a mood which represents events as contrary to reality: at the moment of utterance the subject *you* of (157) was not near the speaker.

Instances of subjunctive I are similar to imperative clauses in that the predicator verb is also in the plain form: we say *I suggest that he start* ... instead of *I suggest that he starts* ... This is their main formal characteristic. (In spite of this formal similarity, instances of subjunctive I are easily distinguishable from imperatives: in contrast to the latter, the former contain overt subjects like *he* (of 156).)

With regard to subjunctive II, we need to distinguish between the **present subjunctive II** and the **past subjunctive II**. Compare, for example, sentences (157) / (158) and (159).

(158)   I wish the news were better (COCA)
(159)   I wish the news had been better (COCA)

The present subjunctive II is a relative tense that conveys simultaneousness: i.e. e.g. in (158) the speaker wishes that the news were better at the moment of utterance. The defining formal characteristic of the present subjunctive II is that with the exception of *be*, all other present subjunctive II wordforms are identical with past tense wordforms. For example, (160) and (161).

(160)   I wish he <u>dumped</u> that bill that we enacted this year that's going to make it a lot worse on you taxpayers to meet your responsibilities (COCA)
(161)   I wish he <u>gave</u> us time to probe (COCA)

With regard to *be*, there is a stylistically relevant free variation between the allowordforms *were* and *was*, the latter being more informal than the former (Huddleston 2002b: 86). For example, (162) and (163)

(162)   I wish I <u>were</u> a better man (COCA)
(163)   I wish I <u>was</u> still there (COCA)

The past subjunctive II, exemplified by the clause *I wish the news had been better*, is a relative tense that conveys anteriority. That is, the speaker wishes that the news had been better at some point in the past prior to the moment of utterance. From a formal point of view, past subjunctive II wordforms like *had been better* of (159) are identical with perfect wordforms of indicative past perfect clauses like *The news had been better*.

This textbook departs from the above named authors in that it regards subjunctive I wordforms like *(he) start* of (156) as allowordforms of indicative mood wordforms like *(he) starts* of e.g. *He starts fighting Titan*. The main justification for this analysis is the non-obligatoriness of subjunctive I in Present-day English. Already in 1964, Ganshina and Vasilevskaya noted that "in Modern English subjunctive I is rapidly falling into disuse" (1964: 204). Similarly, in a more recent reference grammar of English, Quirk et al. (1985: 155) point out that "the subjunctive in modern English is generally an optional and stylistically somewhat marked variant of other constructions [...]". Finally, according to Huddleston and Pullum,

> subjunctives occur as main clauses in a few more or less fixed expressions, as in *God bless you, Long live the Emperor*, etc. Their most common use is as subordinate clauses of the kind [*It's essential that he keep us informed*]. Structurally these differ only in the verb inflection from subordinate clauses with a primary verb-form – and many speakers would here use a present tense in preference to the slightly more formal subjunctive: *It's essential that he keeps us informed*. (Huddleston and Pullum 2005: 32)

To sum it up: subjunctive I wordforms occur in a relatively small number of constructions (mainly subordinate clauses like that of (156)), in which they are often replaceable by indicative wordforms: i.e. instead of saying *I would suggest that he start helping the people of the United States*, a speaker of English can also say *I would suggest that he starts helping the people of the United States*. Given this fact, it is clear that the traditional analysis of subjunctive I as a distinct member of the mood grammeme in English is untenable. As stated above, the plain form *(he) start* of (156) is not a subjunctive I wordform but a stylistically more formal allowordform of the indicative *(he) starts*.

6.5.7 Person

PERSON is the grammatical category which is concerned with the relation between the participants of a **speech event** (i.e. an event that involves the uttering of at least one clause) and the participants of an event denoted by that clause. As an illustration, let us consider (164) and (165).

(164)   *I trust you completely* (COCA)
(165)   *She is a beautiful child* (COCA)

The two participants of any speech event are the **speaker** and the **addressee**. For example, the two participants of the speech event (164) are the person who uttered the clause *I trust you completely* (the speaker) and the person(s) addressed by him or her (his or her addressee(s)). Similarly, the two participants of the speech event (165) are the person who uttered *She is a beautiful child* and the person(s) addressed by him or her. (Normally, the speaker and the addressee are different people. However, in the case of **self-speech**, the speaker is simultaneously the addressee.)

Now, as regards the relation between the participants of the speech events (164) and (165) and the participants of the events denoted by these clauses, one can easily notice that in (164) the speaker and the addressee are simultaneously the two participants of the event denoted by this clause. That is, the speaker is the person who completely trusts the addressee(s). By contrast, the participant of the event expressed by (165) is a person who is neither the speaker nor the addressee. Neither the former nor the latter is a beautiful child.

The category of person can thus be defined as the set of the following mutually exclusive grammatical meanings:

1. '**the first person**', i.e. the speaker
2. '**the second person**', i.e. the addressee
3. '**the third person**', i.e. a person or an object who is neither the speaker nor the addressee

The first person gram in English is the personal pronoun *I*; the second person gram is the personal pronoun *you*; the third person grams are the personal pronouns *he*, *she*, and *it* (differing with regard to gender). Each of these grams has plural wordforms: *I* → *we*, *you* → *you*, *he / she / it* → *they*. In this connection, it is important to observe that while the plural *you* can be semantically analyzed as 'more than one addressee' (e.g. the speaker of (164) completely trusts more than one person), the plural *we* does not mean 'more than one speaker' but 'a group of people which includes the speaker'. If, for example, we change (164) into *We completely trust you*, this sentence will acquire the following meaning: 'the speaker believes that he or she is a member of some group each of whose members, including the speaker, trusts the addressee(s)'. *We* is thus a quasi-idiom in relation to *I*: the signified of the former does not only contain the meanings 'the first person' and 'the plural number' but also the idiomatic meaning 'a group including [the speaker and at least one other individual]'. As for *they*, the idiomatic meaning 'group' is not always part of its signified. For example, if we change (164) in *I trust them completely*, this will not necessarily imply that the speaker conceptualizes the people referred to as *them* as members of some group. These can only be people whom he or she happens to know. Consequently, the signified 'they' can be analyzed in a similar way as the signified of the plural *you*: 'they' = 'more than one individual (excluding both the speaker and the addressee)'.

According to Huddleston and Pullum (2005: 100-101), personal pronouns can fill the same syntactic positions as nouns and, accordingly, must be regarded as a subclass of nouns rather than as an independent syntactic category. Consider, for example, (166).

(166)   *He gave it me!* (COCA)

With the exception of the verb *gave*, which functions as the predicator of the predicate VP *gave it me*, all other syntactic positions in this clause are filled by pronouns: the nominative *he* functions as the subject; the accusative *it* is the indirect object and the accusative *me* is the direct object. (Note that even though the direct object *me* of (166) is semantically very similar to the indirect object *the President* of *People gave the President the right information*, *me* of (166) is nevertheless the direct object: we are justified in claiming this because *me* is placed after another object.[17]) As for the complement position, recall clauses like *it is I / me who ...* in which the complement position can be filled by personal pronouns in either the nominative or the accusative case.

The two major differences between nouns and personal pronouns are as follows.

- In contrast to nouns, personal pronouns cannot be modified by determiners. E.g. we can say *the guy* but not **the he*.

- In contrast to nouns, personal pronouns have only grammatical meanings.

With regard to the latter, Plungian (2000: 255) argues that personal pronouns are **"fully grammaticalized lexemes"**. Indeed, the pronoun *I* carries only the grammatical meanings 'the first person', 'the singular number', and 'the nominative case'. Similarly, the pronoun *me* carries only the grammatical

---

[17] In clauses like (166) where the direct and the indirect object position is expressed by personal pronouns, the order of objects can be reversed. That is, the clause *He gave me it* is as grammatical as *He gave it me*. Nevertheless, the syntactic status of the personal pronouns *me* and *it* must be determined only on the basis of their position in relation to each other. That is, in *He gave me it*, *me* is the indirect object and *it* is the direct object. By contrast, in *He gave it me*, *it* is the indirect object and *me* is the direct object.

meanings 'the first person', 'the singular number', and 'the accusative case'. Perhaps only the quasi-idiomatic *we* can be analyzed as a carrier of the lexical meaning 'a group including [the speaker and at least one other individual]'.

Finally, it must be observed that in addition to personal pronouns, the grammatical category of person has another distinct marker: the agreement between the subject and the predicator verb in the present simple tense. This is especially obvious in the case of the verb *be*, which has three distinct person wordforms: *am* for the first person (singular), *are* for the second person (singular), and *is* for the third person (singular). Also in the past simple tense, we find the distinction between the wordform *was*, which is used with first person and third person singular subjects (i.e. *I was* and *she was*), and the wordform *were*, which is used with second person singular subjects (i.e. *you were*). With the exception of modal verbs like *can*, *may*, *must*, etc., all other verbs distinguish only between present tense first person / second person singular wordforms like *read* (*I read* and *you read*) and present tense third person singular wordforms like *reads* (e.g. *She reads*).

6.5.8 Number

In 6.2 the category of number was used as an illustrative example of a semantic grammeme. As we said, there is a very clear referential difference between one representative of a particular class of objects (e.g. *a book*) and more than one representative of the same class of objects (e.g. *two books*). Given this difference, we can define NUMBER as a grammeme concerned with the expression of the referential contrast '**oneness–more-than-oneness**'.

With regard to number, English nominal lexemes are traditionally classified into the following three categories:

1. **lexemes that can be realized by both singular and plural wordforms**: e.g. BOOK ← *book* and *books*; DEER ← *deer* and *deer*, where the output plural wordform *deer* is identical with the singular input wordform *deer*; etc.

2. **singularia tantum**, i.e. lexemes that can be realized by singular wordforms only: e.g. PERSEVERANCE ← *perseverance*, not **perseverances*; LINGUISTICS ← *linguistics*, not **linguisticses*; etc.

3. **pluralia tantum**, i.e. lexemes that can be realized by plural wordforms only: e.g. TROUSERS ← *trousers*, not **trouser*; POLICE ← *police were investigating*, but not **police was investigating*; etc.

Inflectional morphology

Note that -*s* of *linguistics* is not a plural morph but a quasi-linguistic unit. This analysis is supported by the following facts:

- When the NP *linguistics* fills the subject position, it can only be followed by singular verbal wordforms: i.e. e.g. we can say *Linguistics is the study of language*, but not **Linguistics are the study of language*.

- *Linguistics* is not associated with the meaning 'more-than-oneness': there is only one department of study which we call linguistics.

The existence of singularia and pluralia tantum may raise the question of whether NUMBER fulfills the obligatoriness requirement which we discussed in 6.1. That is, if there are nominal lexemes that lack either singular or plural wordforms, can we still analyze the singular–plural contrast as an obligatory grammatical contrast? The answer to this question is 'yes'. Both singularia and pluralia tantum fulfill the obligatoriness requirement because both of them are associated with one of the two meanings forming the grammatical category NUMBER: singularia tantum express the meaning 'singularity' and pluralia tantum express the meaning 'plurality'. There are no nouns expressing neither the meaning 'singularity' nor the meaning 'plurality'.

In addition to this, observe that the existence of both singularia and pluralia tantum is motivated by inherent properties of the concepts which they denote. For example, both *perseverance* and *linguistics* are **uncountable nouns**, i.e. nouns that cannot be modified by **cardinal numerals** like *one*, *two*, *three*, etc. (Huddleston and Pullum 2005: 86). That is, we cannot say **two perseverances* and **two linguisticses*. The obvious explanation for this is that there is only one type of behavior that we conceptualize as perseverance – i.e. a behavior characterized by 'continued effort to do or achieve something despite difficulties, failure, or opposition' (MWO) – and there is only one department of study which we conceptualize as linguistics. Similarly, the plural nature of both *trousers* and *police* can be attributed to the 'plural' properties of the objects 'trousers' and 'police' which they denote. The latter is not really an object but an institution consisting of numerous police officers. And *trousers* is an example of what Quirk et al. (1985: 300) call **summation plurals**, i.e. nouns denoting 'tools, instruments, and articles consisting of two equal parts which are joined together'. Similar pluralia tantum include *glasses*, *binoculars*, *scissors*, *tongs*, *tweezers*, etc.

Many nominal vocables consist of both countable and uncountable lexemes. For example, the vocable COFFEE consists of the uncountable $COFFEE_1$ (e.g. *a cup of coffee*) and the countable $COFFEE_2$ (e.g. *two coffees*). The latter is a product of quasi-idiomatization of the former: i.e. while $COFFEE_1$ means 'coffee, i.e. a beverage made by percolation, infusion, or decoction from the roasted and ground seeds of a coffee plant' (MWO), COFFEE₂ means 'a cup of coffee'. In other words, the signified of COFFEE₂ contains not only the signified 'coffee', inherent in the component *coffee*, but also the idiomatic meaning 'a cup of'. As Quirk et al. (1985: 298-299) observe, quasi-idiomatic countable lexemes like COFFEE₂ are typically products of semantic change of corresponding input uncountable lexemes. Apart from COFFEE₂, this is true of e.g. BEER₂ 'a glass of beer' (e.g. *two beers*) and PLEASURE₂ 'an instance of pleasure' (e.g. *the pleasures of life*). The reverse situation is, however, possible as well. That is, a countable input lexeme can give rise to an uncountable output lexeme. For example, Radden (2006) discusses the uncountable uses of *house* in sentences like (167).

(167)   How much <u>house</u> can you afford? (http://tinyurl.com/4qecz)

In this sentence *house* does not refer to a house as a countable object but denotes the substance 'house', just as the uncountable *coffee* of e.g. *a cup of coffee* denotes the substance 'coffee'. The semantic change HOUSE₁ 'a countable object' → HOUSE₂ 'house as substance' is a relatively recent instance of semantic change in English, so that sentences like (167) may still be perceived as ungrammatical by some English speakers.

6.5.9 Degrees of comparison

As defined in Section 6.1, DEGREES OF COMPARISON or simply GRADE is the set of the mutually exclusive meanings 'the positive degree of comparison', 'the comparative degree of comparison', and 'the superlative degree of comparison'. From a formal standpoint, a positive adjectival wordform like *pretty* and a positive adverbial wordform like *soon* can be seen as input wordforms that are used for producing both comparative and superlative adjectival and adverbial wordforms: the comparative *prettier / sooner* and the superlative *prettiest / soonest* can be analyzed as products of inflectional affixation of the input positive wordforms *pretty / soon* by means of the suffixes *-er* and *-est*. Likewise, the analytic comparatives *more beautiful / more beautifully* and the analytic superlatives *most beautiful / most beautifully* are products of combining the input positive wordforms *beautiful / beautifully* with the analytic grams *more* and *most*.

By contrast, from a semantic standpoint, a positive wordform can often be considered an output wordform in relation to a corresponding input comparative wordform. Consider, for example, the signified of the positive wordform *pretty* in an NP like *a pretty woman*. Evidently, the characterization of a woman as a pretty woman is based on a comparison of that woman with other women who are less pretty than her. As was argued by Sapir (1944: 95), "all comparatives are primary in relation to their corresponding absolutes ('positives')". Thus, according to Sapir, *large* means 'larger than of average size' and *small* means 'smaller than of average size'. Similarly, we can argue that a pretty woman is prettier than the average woman; a big city is bigger than the average city; a difficult task is more difficult than the average task; etc. What this means is that the formally less complex positive wordforms *large, small, pretty, big, difficult* are semantically more complex than the formally more complex comparative wordforms *larger, smaller, prettier, bigger, more difficult*: the signifieds of the former can be analyzed as quasi-idioms in relation to the signifieds of the latter. For example, the signified 'large' contains the signified 'larger than', inherent in the input comparative wordform *larger*, plus the idiomatic meaning 'average'. (The latter can be regarded as an idiomatic meaning because, according to Sapir, the signifieds of some positive wordforms do not contain this meaning. For instance, *good* is analyzed by Sapir not as 'better than the average quality' but as 'better than indifferent'. Similarly, *much* is said to mean not 'more than the average amount' but 'more than a fair amount'.)

Wierzbicka (1972: 71-92) revises Sapir's claim that all comparative wordforms are semantically primary in relation to their corresponding positive wordforms. According to her, in the case of e.g. *sick* and *healthy*, the positive meanings cannot be analyzed as 'sicker / healthier than the average person': in contrast to a pretty woman, the characterization of a person as sick or healthy does not seem to be based on a comparison of that person with other people who are less sick and less healthy than that person. Hence, as Wierzbicka concludes, *sick* and *healthy* can be regarded as positives that are semantically less complex than the corresponding comparatives *sicker* and *healthier*: a sicker person is a person who is more sick than some other sick person and a healthier person is a person who is more healthy than some other healthy person.

Another important revision made by Wierzbicka concerns the meaning 'average', which, according to Sapir, is part of the signifieds of positive wordforms like *large, small, pretty, big,* and *difficult*. As she points out,

> it is clear what *average* means when used in respect of the elements of a finite set. When used, however, in respect of an open, infinite set, it becomes a problem in itself. (Wierzbicka 1972: 73; italics mine)

Indeed, what precisely is meant by *the average size, the average woman, the average city, the average task,* etc.? Given this obvious difficulty, Wierzbicka proposes the following formulaic solution:

- X_1 is small. = X_1 is a small X. = X_1 is smaller than one would expect an X to be. (Wierzbicka 1972: 73)

In other words, we have a certain expectation about the size of representatives of a particular class of objects. One representative of this class turns out to be smaller than we thought a representative of this class would be. Similarly, a pretty woman is not prettier than the average woman but a woman prettier than one would expect a woman to be.

As for the superlative degree of comparison, the semantic analysis of a superlative wordform depends on whether it realizes the **relative** or the **absolute superlative** (Sapir 1944: 113). For example, the superlative wordform *prettiest* of e.g. *the prettiest of the three women* is an instance of the relative superlative: it means 'prettier than two other women (of the three women the speaker has in mind; it is possible that some other woman is prettier than her.)'. By contrast, *prettiest* of *the prettiest woman* is an instance of the absolute superlative: it means 'prettier than all other women'.

In 6.5.8 we established that not all nominal lexemes have both singular and plural wordforms. Similar to this, there are adjectival and adverbial lexemes that lack the comparative and the superlative wordforms. For example, an American citizen cannot be **a more American citizen* than other American citizens and he or she could not **more always* live in New York. The adjective *American* is thus a **non-gradable adjective** and *always* is a **non-gradable adverb**.

As in the case of uncountable nouns, linguistic non-gradability is motivated by the extra-linguistic non-gradability of concepts denoted by non-gradable adjectives and adverbs. That is, for example, a person can either be an American or a non-American citizen: there are no more or less American citizens. Likewise, there can only be the distinction between always and not always living in New York, but not between more and less always living there. Also, as in the case of nominal vocables like BEER, COFFEE, PLEASURE, HOUSE, etc., which consist of both countable and uncountable lexemes, there are many adjectival vocables containing both gradable and non-gradable lexemes. For example, an American citizen cannot be **more American citizen* than other American citizens, but he or she can be *more American* (i.e. have more American qualities) than other Americans. Similarly, there can be no **more public transport*, but there can be *a more public discussion*.

6.5.10 Numerical qualification

Finally, let us discuss the grammatical category NUMERICAL QUALIFICATION. Following Khaimovich and Rogovskaya (1967: 93), this grammeme can be defined as the set of the mutually exclusive grammatical meanings '**numerical quantity**' and '**numerical order**'. The former meaning is expressed by **cardinal numerals** like *one, two, three, four*, etc. The latter meaning is expressed by **ordinal numerals** like *first, second, third, fourth*, etc.

Cardinal numerals respond to the question *how many?* (OED). For example, (168).

(168)   Look again . *How many Enigma days?' 'Three, five, two, seven... and the only sample taken was by me on the day we arrived!'* (COCA)

Ordinal numerals, by contrast, "designate the place [...] occupied by an item in an ordered sequence" (MWO). For example, in the ordered sequence SIX CHAPTERS OF THE PRESENT TEXTBOOK ON ENGLISH MORPHOLOGY, the present chapter on inflectional morphology occupies the sixth place.

Sometimes cardinal numerals are used for the expression of ordinal meanings. For example, instead of saying *the sixth chapter*, we can also say *chapter six*. In the latter variant, the cardinal numeral *six* does not answer the question *how many chapters?* but designates the sixth place occupied by some chapter in the ordered sequence CHAPTERS OF SOME BOOK. This is clearly an idiomatic use of cardinal numerals.

As was mentioned in 6.3.1, ordinal numerals are formed in English with the help of the suffix *-th*. This suffix attaches to input cardinal wordforms: e.g. *seventh ← seven + -th*. The suffix *-th* is a very productive inflectional suffix that attaches to the majority of input cardinals:

- *fourth ← four + -th*
- *ninth ← nine + -th*
- *forty-sixth ← forty-six + -th*

The only exceptions are cardinal numerals *one*, *two*, and *three*, which form ordinal wordforms with the help of suppletion. That is, *first* is the suppletive ordinal wordform of the input cardinal *one*; *second* is the suppletive ordinal wordform of the input cardinal *two*; *third* is the suppletive ordinal wordform of the input cardinal *three*. In addition, suppletion produces ordinal wordforms of compound input cardinals headed by *one*, *two*, and *three*. That is, for example, *twenty-first* is the suppletive ordinal wordform of the compound input cardinal *twenty-one*; *one hundred and fifty-second* is the suppletive ordinal wordform of the input cardinal *one hundred and fifty-two*; *one thousand and ninety-third* is the suppletive ordinal wordform of the input cardinal *one thousand and ninety-three*; etc.

The regular ordinal-forming suffix *-th* has two allomorphs: /θ/ and /əθ/. The latter occurs in ordinal wordforms whose input cardinal wordforms end in *-ty*. For example:

- /ˈsɪk.sti.əθ/ ← /ˈsɪk.sti/ + /əθ/
- /ˈeɪ.ti.əθ/ ← /ˈeɪ.ti/ + /əθ/

- /ˈnaɪn.ti.əθ/ ← /ˈnaɪn.ti/ + /əθ/

In all other cases, the suffix has the realization /θ/. For example:

- /sɪksθ/ ← /sɪks/ + /θ/
- /nʌɪnθ/ ← /nʌɪn/ + /θ/
- /eɪˈtiːnθ/ ← /eɪˈtiːn/ + /θ/

Note also that in the case of *fifth* /fɪfθ/ and *twelfth* /twɛlfθ/, the addition of -*th* is accompanied by apophony. That is, [aɪv] of the input cardinal *five* changes in [ɪf] in *fifth*. Similarly, the final consonant [v] of the input cardinal *twelve* changes in [f] in *twelfth*.

6.6 Exercises

1. Make sure you can explain each of the key terms printed in boldface (ideally, using your own examples).

2. Which of the following statements are true?

a) Grammatical meanings that are members of the same grammeme are mutually exclusive.
b) Semantic grammemes in English are considerably outnumbered by syntactic grammemes.
c) Grammatical apophony usually produces irregular wordforms.
d) *Get*-passives are as obligatory as *be*-passives.
e) English has the prepositional dative case.
f) Simple tenses have two tense loci: the moment of utterance and some other event in the past.
g) The progressive aspect in English is associated with the meaning 'the middle phase of an event'.
h) The subjunctive mood is a non-fact mood which represents an action denoted by a clause as problematic.
i) The personal pronoun *you* expresses the meaning 'the third person'.
j) A nominal vocable can consist of both countable and uncountable lexemes.

3. State which wordform-building mechanisms produced the following wordforms.

a) the genitive third person *its*
b) the passive past tense *was discussed*

c) the past tense *began*
d) the genitive plural *boys'*
e) the nominative plural *boys*
f) the first person plural nominative *we*
g) the ordinal numeral *thirty-fourth*
h) the imperative wordform *do*
i) the comparative wordform *more British*
j) the second person present tense plural *are*

4. Comment on the temporal meanings of the underlined events. State whether they are in the present or in the past tense and whether they are instances of 1) simple or perfect tenses, and 2) absolute or relative tenses.

a) *I came here to <u>see you</u>* (COCA)
b) *<u>When he came home a few months ago</u>, he was a changed man* (COCA)
c) *<u>Today I have had an extraordinary experience</u>* (COCA)
d) *<u>She will not be so eager</u> when she notices me* (COCA)
e) *<u>They go</u> and they flip the pancakes* (COCA)
f) *<u>If he had been kept in an institution</u>, he certainly wouldn't have done so well* (COCA)
g) *Through these kin relationships, participants reported <u>having received emotional and concrete support</u>* (COCA)
h) *We did sleep together, <u>we had done so many times over the years</u>, but not in a sexual way* (COCA)
i) *[...] it's time <u>we did something about crime</u>* (COCA)
j) *I'm sorry to <u>have come here uninvited</u> [...]* (COCA)

5. Prove that:

a) the clause *They kissed flauntingly* (COCA) denotes a dynamic event;
b) the clause *[...] Clinton was far from a passive observer* (COCA) denotes a state;
c) the clause *Another car crashed into ours* (COCA) denotes an achievement;
d) the clause *He was lying in an awkward position* (COCA) denotes a process;
e) the clause *She came to me* (COCA) denotes a telic event;
f) the clause *She has stayed in Grogan's Mill for the past two decades* (COCA) denotes an atelic event;
g) the clause *Sleep well* (COCA) is in the imperative mood;
h) the underlined clause in *Her parole requires that <u>she stay in Peru</u> until her sentence ends in November 2015* (COCA) is an instance of subjunctive I;
i) the clause *She is a feature at every party* (COCA) is in the indicative mood;

j) the underlined clause *If she were a different woman, would you think the same of her?* (COCA) is an instance of subjunctive II.

6. Explain the ungrammaticality of the following sentences, i.e. what is wrong with them.

a) *The suspect gave the wrong answer the committee*
b) *Him was frightened by that*
c) *His results were gooder than mine*
d) *He watches TV* meaning 'he is doing this at the moment of utterance'
e) *The she is a pretty woman*
f) *Awake was stayed by me*
g) *Another car continued crashing into ours*
h) *He am a good guy*
i) *Martinez doesn't intimidate easily by other people*
j) *Barack Obama is being the U.S. president*

6.7 Further reading

With the exception of the first two sections, this chapter was largely based on Huddleston and Pullum's *The Cambridge Grammar of the English Language* (2002), a must-read for any student of English theoretical linguistics. Earlier major reference grammars of English are Biber et al. (1999) and Quirk et al. (1985).

An excellent introduction to **grammatical semantics** (i.e. the branch of semantics which is concerned with grammatical meanings like those discussed in this chapter) is Frawley (1992). A more concise introduction can be found in Cruse (2004: 275-311). A more advanced reading is Mel'čuk (2006).

Due to space limitations, this chapter does not provide a discussion of how wordform-building mechanisms in English have changed with the course of time. The reader is therefore referred to the chapters Hogg (1992; especially 122-164) and Lass (1992; especially 91-147) in *The Cambridge History of the English Language* (Volumes I and II). The former discusses wordform-building in Old English; the latter deals with the Middle English period.

Key to exercises

Chapter 1

2. Statements b, d, e, h, i are correct.

3. Words b, c, f, i are complex words.

4. a) predicate, b) subject, c) complement, d) adjunct, e) adjunct, f) object, g) predicate, h) subject, i) object, j) adjunct.

5. Forms a, b, c, d, f, h, i, j are words.

Chapter 2

2. Statements a, f, g are correct.

3. a) contrastive distribution, b) complementary distribution, c) contrastive distribution, d) complementary distribution, e) free variation, f) free variation, g) complementary distribution, h) contrastive distribution, i) complementary distribution, j) complementary distribution.

4. a) *green* is a morfoid, *house* is a morph, b) *table* is a monomorphemic word, c) both *bar* and *man* are morphs, d) both *fore-* and *tell* are morphs, e) both *with* and *in* are submorphs, f) both *per-* and *-mit* are submorphs, g) *great* is a monomorphemic word, h) both *after* and *party* are morphs, i) both *re-* and *do* are morphs, j) *bird* is a morfoid, *brain* is a morph.

5. a) derivational suffix, b) analytic free form / root, c) free form / root, d) inflectional suffix, e) derivational suffix, f) inflectional suffix, g) bound submorph (cf. *progress* and *congress*), h) derivational prefix, i) free morfoid / root, j) free form / root.

Chapter 3

2. Statements a, d, g, j are correct.

3. a) contrastive distribution, b) complementary distribution, c) complementary distribution, d) free variation, e) complementary distribution, f) contrastive distribution, g) complementary distribution, h) free variation, i) free variation, j) complementary distribution.

4. a) lexes of two different lexemes, b) lexes of two different lexemes which form a lexeme family, c) lexes of two different lexemes, d) wordforms of the same lexeme, e) lexes of two different lexemes which form a vocable, f) lexes of two different lexemes which form a lexeme family, g) lexes of two different lexemes which form a vocable, h) lexes of two different lexemes which form a lexeme family, i) wordforms of the same lexeme, j) lexes of two different lexemes.

5. a) compound / semi-idiom, b) simplex, c) compound / quasi-idiom, d) prefixed word / isomorphic word, e) compound / quasi-idiom, f) simplex, g) suffixed word / quasi-idiom, h) compound / quasi-idiom, i) prefixed word / isomorphic word, j) compound / semi-idiom.

Chapter 4

2. Statements b, d, e, f, h, i, j are correct.

3. a) blending, b) anisomorphic affixation, c) morphological conversion, d) morphological conversion, e) compounding, f) blending, g) anisomorphic affixation, h) full-idiomatization of the NP *mission from God*, i) compounding / semi-idiomatization of the NP *soft power*, j) morphological conversion.

4. a) apophony, b) abbreviation, c) clipping, d) affixation, e) clipping, f) syntactics' change, g) suppletion, h) clipping, i) abbreviation, j) suppletion.

5. Suffixes a, b, d, f, g, h, i are productive.

Chapter 5

2. Statements a, d, e, g, i, j are correct.
3. 5
4. German
5. no
6. 11

Key to exercises

7. 3
8. between 1900-1949
9. 10

Chapter 6

2. Statements a, c, g, h, j are correct.

3. a) affixation, b) analytic formation + affixation, c) apophony, d) signifier-sharing with *boys*, e) affixation, f) suppletion, g) affixation, h) signifier-sharing with the plain form / present tense *do*, i) analytic formation, j) suppletion.

4. a) indefinite infinitive / relative tense, b) past simple / absolute tense, c) present perfect / absolute tense, d) present simple / absolute tense, e) present simple / absolute tense, f) perfect subjunctive II / relative tense, g) perfect gerund / relative tense, h) past perfect / absolute tense, i) present subjunctive II / relative tense, j) perfect infinitive / relative tense.

5. a) progressivization possible, b) progressivization impossible, c) cannot combine with e.g. *begin*, d) is in the progressive aspect, e) it was begun in order to be finished, f) contains a duration adjunct, g) the verb is in the plain form / no overt subject, h) the verb is in the plain form / an overt subject, i) the predicator verb agrees with the subject in person and number, j) the predicator position is filled by *were*.

6. a) the direct object precedes the indirect object, b) the subject is expressed by a personal pronoun in the accusative case, c) *good* does not have an inflectional comparative wordform, d) present simple tense does not combine freely with processes like watching TV, e) personal pronouns do not take determiners, f) complements cannot serve as subjects of associated passive clauses, g) the clause denotes an achievement, h) no person agreement between the subject and the predicator, i) middle clauses suppress the *by*-phrase, j) the clause denotes a state.

References

Algeo, J. (1998): "Vocabulary", *The Cambridge History of the English Language, Volume IV 1776-1997*, ed. S. Romaine. Cambridge: Cambridge University Press. 57-91.
Anderson, S. (1992): *A-Morphous Morphology*. Cambridge: Cambridge University Press.
Apresjan, Y. (1974): *Leksičeskaja Semantika. Sinonimičeskie Sredstva Jazyka* [*Lexical Semantics. Synonymic Means of a Language*]. Moscow: Nauka.
Aronoff, M. (1976): *Word Formation in Generative Grammar*. Cambridge, MA: MIT Press.
Aronoff, M. & K. Fudeman (2005): *What is Morphology?* Malden: Blackwell.
Ayto, J., J. Siefring & J. Speake (eds.) (2009): *Oxford Dictionary of English Idioms*. Third edition. Oxford: Oxford University Press.
Balteiro, I. (2007): *The Directionality of Conversion in English. A Dia-Synchronic Study*. Bern: Peter Lang.
Bauer, L. (1979): "On the need for pragmatics in the study of nominal compounding", *Journal of Pragmatics* 3, 45-50.
Bauer, L. (2003): *Introducing Linguistic Morphology*. Second edition. Edinburgh: Edinburgh University Press.
Bauer, L. (2005): "Productivity: theories", *Handbook of Word-Formation*, ed. P. Štekauer & R. Lieber. Dordrecht: Springer. 315-332.
Bauer, L. & R. Huddleston (2002): "Lexical word-formation", *The Cambridge Grammar of the English Language*, R. Huddleston & G. Pullum. Cambridge: Cambridge University Press. 1621-1721.
Benczes, R. (2006): *Creative Compounding in English. The Semantics of Metaphorical and Metonymical Noun-Noun Combinations*. Amsterdam: John Benjamins.
Biber, D., S. Johansson, G. Leech, S. Conrad & E. Finegan (1999): *Longman Grammar of Spoken and Written English*. Harlow: Longman.
Blake, B. (1994): *Case*. Cambridge: Cambridge University Press.
Bloomfield, L. (1973[1934]): *Language*. Reprint. London: George Allen & Unwin LTD.
Burger, H., D. Dobrovol'skij, P. Kühn & N. Norrick (eds). (2007): *Phraseologie. Ein Internationales Handbuch der Zeitgenössischen Forschung*. Berlin: Walter de Gruyter.
Burnley, D. (1992): "Lexis and semantics", *The Cambridge History of the English Language, Volume II 1066-1476*, ed. N. Blake. Cambridge: Cambridge University Press. 409-499.
Comrie, B. (1976): *Aspect*. Cambridge: Cambridge University Press.

Cruse, A. (2004): *Meaning in Language. An Introduction to Semantics and Pragmatics.* Second edition. Oxford: Oxford University Press.

Cruttenden, A. (2008): *Gimson's Pronunciation of English.* Seventh edition. London: Hodder Education.

Davies, M. (2004-): *BYU-BNC: The British National Corpus.* Available online at <http://corpus.byu.edu/bnc>, accessed 29 November 2011.

Davies, M. (2008-): *The Corpus of Contemporary American English* (COCA). Available online at <http://www.americancorpus.org>, accessed 29 November 2011.

Dobrovol'skij, D. & E. Piirainen (2005): *Figurative Language. Cross-Cultural and Cross-Linguistic Perspectives.* Bingley: Emerald.

Dobrovol'skij, D. & E. Piirainen (2009): *Zur Theorie der Phraseologie. Kognitive und Kulturelle Aspekte.* Tübingen: Stauffenburg.

Fiedler, S. (2007): *English Phraseology. A Coursebook.* Tübingen: Gunter Narr.

Finegan, E. (2004): *Language. Its Structure and Use.* Fourth edition. Boston: Thomson Wadsworth.

Frawley, W. (1992): *Linguistic Semantics.* Hilsdale: Lawrence Erlbaum.

The Free Dictionary. Available at <http://www.thefreedictionary.com/>, accessed 15 October 2011.

Ganshina, M. & N. Vasilevskaya (1964): *English Grammar.* Ninth revised edition. Moscow: Vysšaja škola.

Giegerich, H. (2009): "Compounding and lexicalism", *The Oxford Handbook of Compounding*, ed. R. Lieber & P. Štekauer. Oxford: Oxford University Press. 178-200.

Ginzburg, R, S. Khidekel, G. Knyazeva & A. Sankin (1979): *A Course in Modern English Lexicology.* Second edition. Moscow: Vysšaja škola.

Gut, U. (2009): *Introduction to English Phonetics and Phonology.* Frankfurt am Main: Peter Lang.

Haan, F. (2006): "Typological approaches to modality", *The Expression of Modality*, ed. W. Frawley. Berlin: Mouton de Gruyter. 27-69.

Halle, M. (2005): "Palatalization/velar softening: What it is and what it tells us about the nature of language", *Linguistic Inquiry* 36/1, 23-41.

Hanks P, K. Hardcastle & F. Hodges (2006): *A Dictionary of First Names.* Second edition. Oxford: Oxford University Press.

Harris, Z. (1942): "Morpheme alternants in linguistic analysis", *Language* 18/3, 169-180.

Harris, Z. (1945): "Discontinuous morphemes", *Language* 21/3, 121-127.

Haspelmath, M. (2011): "The indeterminacy of word segmentation and the nature of morphology and syntax", *Folia Linguistica* 45/1, 31-80.

Haspelmath, M. & A. Sims (2010): *Understanding Morphology.* Second edition. London: Hodder Education.

Heringer, H.-J. (2009): *Morphologie.* Paderborn: Wilhelm Fink.

Hockett, C. (1947): "Problems of morphemic analysis", *Language* 23/4, 321-343.
Hogg, R. (1992): "Phonology and morphology", *The Cambridge History of the English Language, Volume I The Beginnings to 1066*, ed. R. Hogg. Cambridge: Cambridge University Press. 67-167.
Hohenhaus, P. (2005): "Lexicalization and institutionalization", *Handbook of Word-Formation*, ed. P. Štekauer & R. Lieber. Dordrecht: Springer. 353-373.
Holder, R. (2008): *Oxford Dictionary of Euphemisms*. Fourth edition. Oxford: Oxford University Press.
Horn, L. & G. Ward (eds.) (2004): *The Handbook of Pragmatics*. Malden: Blackwell Publishing.
Huddleston, R. (2002a): "The verb", *The Cambridge Grammar of the English Language*, R. Huddleston & G. Pullum. Cambridge: Cambridge University Press. 71-212.
Huddleston, R. (2002b): "The clause: complements", *The Cambridge Grammar of the English Language*, R. Huddleston & G. Pullum. Cambridge: Cambridge University Press. 213-321.
Huddleston, R. & G. Pullum (2002): *The Cambridge Grammar of the English Language*. Cambridge: Cambridge University Press.
Huddleston, R. & G. Pullum (2005): *A Student's Introduction to English Grammar*. Cambridge: Cambridge University Press.
Jackendoff, R. (2009): "Compounding in the parallel architecture and conceptual semantics", *The Oxford Handbook of Compounding*, ed. R. Lieber & P. Štekauer. Oxford: Oxford University Press. 105-128.
Jakobson, R. (1959): "Boas's view of grammatical meaning", *Selected Writings. Volume II*, R. Jakobson (1971). Berlin/New York: Mouton de Gruyter. 489-496.
Jakobson, R. (1972): "Some questions of meaning", *On Language*, R. Jakobson & K. Pomorska (1990), ed. L. Waugh & M. Monville-Burston. Cambridge, MA: Harvard University Press. 315-324.
Kaplan, J. (1995): *English Grammar. Principles and Facts*. Englewood Cliffs: New Jersey.
Kastovsky, D. (1992): "Semantics and vocabulary", *The Cambridge History of the English Language, Volume I The Beginnings to 1066*, ed. R. Hogg. Cambridge: Cambridge University Press. 290-408.
Katamba, F. & J. Stonham (2006): *Morphology*. Second edition. Basingstoke: Palgrave Macmillan.
Keller, R. & I. Kirschbaum (2003): *Bedeutungswandel. Eine Einführung*. Berlin: Walter de Gruyter.
Khaimovich, B. & B. Rogovskaya (1967): *A Course in English Grammar*. Moscow: Vysšaja škola.

Kluge, F. (2002): *Etymologisches Wörterbuch der Deutschen Sprache.* 24th edition. Berlin: Walter de Gruyter.
Kövecses, Z. & G. Radden (1998): "Metonymy: developing a cognitive linguistic view", *Cognitive Linguistics* 9/1, 37–77.
Kreyer, R. (2010): *Introduction to English Syntax.* Frankfurt am Main: Peter Lang.
Lakoff, G. (1987): *Women, Fire, and Dangerous Things. What Categories Reveal about the Mind.* Chicago: The University of Chicago Press.
Lakoff, G. & M. Johnson (1980): *Metaphors We Live by.* Chicago: The University of Chicago Press.
Lakoff, G. & M. Johnson (1999): *Philosophy in the Flesh. The Embodied Mind and its Challenge to Western Thought.* New York: Basic Books.
Lass, R. (1992): "Phonology and morphology", *The Cambridge History of the English Language, Volume II 1066-1476*, ed. N. Blake. Cambridge: Cambridge University Press. 23-155.
Lieber, R. (2009): "IE, Germanic: English", *The Oxford Handbook of Compounding*, ed. R. Lieber & P. Štekauer. Oxford: Oxford University Press. 357-369.
Lieber, R. (2010): *Introducing Morphology.* Cambridge: Cambridge University Press.
Lieber, R. & P. Štekauer (2009a): "Introduction: status and definition of compounding", *The Oxford Handbook of Compounding*, ed. R. Lieber & P. Štekauer. Oxford: Oxford University Press. 3-18.
Lieber, R. & P. Štekauer (eds.) (2009b): *The Oxford Handbook of Compounding.* Oxford: Oxford University Press.
Lipka, L. (2002): *English Lexicology. Lexical Structure, Word Semantics and Word-Formation.* Third edition. Tübingen: Gunter Narr.
Löbner, S. (2002): *Understanding Semantics.* London: Arnold.
Lyons, J. (1968): *Introduction to Theoretical Linguistics.* Cambridge: Cambridge University Press.
Marchand, H. (1969). *The Categories and Types of Present-Day English Word-Formation.* Second completely revised and enlarged edition. Munich: Beck.
Maslov, Y. (1973): "Universal'nye semantičeskie komponenty v soderžanii grammatičeskoj kategorii soveršennogo/nesoveršennogo vida [Universal components in the content of the grammatical category perfective/imperfective aspect]", *Izbrannye Trudy. Aspektologija. Obščee Jazykoznanie [Selected Works. Aspectology. General Linguistics]*, Y. Maslov (2004). Moscow: Jaziki Slavjanskoj Kultury. 396-410.
Maslov, Y. (1978): "K osnovanijam sopostavitel'noj aspektologii [To the bases of comparative aspectology]", *Izbrannye Trudy. Aspektologija. Obščee Jazykoznanie [Selected Works. Aspectology. General Linguistics]*, Y. Maslov (2004). Moscow: Jaziki Slavjanskoj Kultury. 305-364.

McMahon, A. (2002): *An Introduction to English Phonology*. Edinburgh: Edinburgh University Press.

Mel'čuk, I. (1968): "The structure of linguistic signs and formal-semantic relations between them", *The Russian Language in the Meaning-Text Perspective*, I. Mel'čuk (1995). Moscow/Vienna: Wiener Slawistischer Almanach. 425-441.

Mel'čuk, I. (1979): "Syntactic, or lexical, zero", *The Russian Language in the Meaning-Text Perspective*, I. Mel'čuk (1995). Moscow/Vienna: Wiener Slawistischer Almanach. 169-205.

Mel'čuk, I. (1982): *Towards a Language of Linguistics. A System of Formal Notions for Theoretical Morphology*. Munich: Fink.

Mel'čuk, I. (1995): "Phrasemes in language and phraseology in linguistics", *Idioms: Structural and Psychological Perspectives*, ed. M. Everaert. Hilsdale: Lawrence Erlbaum Associates. 167-223.

Mel'čuk, I. (1997): *Cours de Morphologie Générale. Cinquième Partie: Signes Morphologiques*. Montréal: Les Pr. de l'Univ. de Montréal.

Mel'čuk, I. (2001). *Kurs Obščej Morfologii. Morfologičeskie Znaki* [*Course in General Morphology. Morphological Signs*]. Moscow/Vienna: Wiener Slawistischer Almanach.

Mel'čuk, I. (2006): *Aspects of the Theory of Morphology*. Berlin: Mouton de Gruyter.

Mel'čuk, I. & A. Zholkovsky (1988): "An explanatory combinatorial dictionary of Modern Russian", *The Russian Language in the Meaning-Text Perspective*, I. Mel'čuk (1995). Moscow/Vienna: Wiener Slawistischer Almanach. 17-54.

Merriam-Webster Online. Available at <http://www.merriam-webster.com/>, accessed 29 November 2011.

Moskvin, V. (2010): *Èvfemizmy v Leksičeskoj Sisteme Sovremennogo Russkogo Jazyka* [*Euphemisms in the Lexical System of Present-Day Russian*]. Fourth edition. Moscow: Lenand.

Mühleisen, S. (2010): *Heterogeneity in Word-Formation Patterns*. Amsterdam: John Benjamins.

Nida, E. (1948): "The identification of morphemes", *Language* 24/4, 414-441.

Nida, E. (1974): *Morphology. The Descriptive Analysis of Words*. Second edition. Ann Arbor: The University of Michigan Press.

Oxford English Dictionary. Available at <http://oed.com/>, accessed 29 November 2011.

Palmer, F., R. Huddleston & G. Pullum (2002): "Inflectional morphology and related matters", *The Cambridge Grammar of the English Language*, R. Huddleston & G. Pullum. Cambridge: Cambridge University Press. 1565-1619.

Payne, J. & R. Huddleston (2002): "Nouns and noun phrases", *The Cambridge Grammar of the English Language*, R. Huddleston & G. Pullum. Cambridge: Cambridge University Press. 323-523.

The Phrase Finder. Available at <http://www.phrases.org.uk/>, accessed 29 November 2011.

Plag, I. (2003). *Word-Formation in English*. Cambridge: Cambridge University Press.

Plungian, V. (2000): *Obščaja Morfologija [General Morphology]*. Moscow: EditorialURSS.

Pullum, G. & R. Huddleston (2002): "Adjectives and adverbs", *The Cambridge Grammar of the English Language*, R. Huddleston & G. Pullum. Cambridge: Cambridge University Press. 525-595.

Quirk, R., S. Greenbaum, G. Leech & J. Svartvik (1985): *A Comprehensive Grammar of the English Language*. London/New York: Longman.

Radden, G. (2006): "Metonymic blending and the construction of meaning", Presentation held at the Second International Workshop on *Metaphor and Discourse*, University of Jaume I de Castelló, February 2-3 2006.

Rainer, F. (2005): "Constrains on productivity", *Handbook of Word-Formation*, ed. P. Štekauer & R. Lieber. Dordrecht: Springer. 335-352.

Sapir, E. (1944): "Grading, a study in semantics", *Philosophy of Science* 11/2, 93-116.

Saussure, F. (1973): *Cours de Linguistique Générale*. Paris: Payot.

Schmid, H.-J. (2008): "New words in the mind: Concept-formation and entrenchment of neologisms", *Anglia - Zeitschrift für englische Philologie* 126/1, 1-36.

Searle, J. (1975): "Indirect speech acts", *Speech Acts. Syntax and Semantics. Volume 3*, ed. P. Cole & J. Morgan. New York: Academic Press. 59-82.

Stein, D. & S. Wright (eds.) (1995). *Subjectivity and Subjectivisation*. Cambridge: Cambridge University Press.

Štekauer, P. & R. Lieber (eds.) (2005). *Handbook of Word-Formation*. Dordrecht: Springer.

Taylor, J. (2002): *Cognitive Grammar*. Oxford: Oxford University Press.

Tokar, A. (2009): *Metaphors of the Web 2.0. With Special Emphasis on Social Networks and Folksonomies*. Frankfurt am Main: Peter Lang.

Trubetzkoy, N. (1939): *Grundzüge der Phonologie*. Prag: Travaux du Cercle Linguistique de Prague.

Trubetzkoy, N. (1969): *Principles of Phonology*. Berkeley: University of California Press.

Ullmann, S. (1957): *The Principles of Semantics*. Oxford: Basil Blackwell.

Urban Dictionary. Available at <http://www.urbandictionary.com/>, accessed 29 November 2011.

Ward, G., B. Birner & R. Huddleston (2002): "Information packaging", *The Cambridge Grammar of the English Language*, R. Huddleston & G. Pullum. Cambridge: Cambridge University Press. 1363-1447.

Wierzbicka, A. (1972): *Semantic Primitives*. Frankfurt am Main: Athenäum.

Wikipedia. Available at <http://en.wikipedia.org/wiki/Main_Page>, accessed 29 November 2011.

WordNet. Available at <http://wordnetweb.princeton.edu/perl/webwn>, accessed 29 November 2011.

Word Spy. Available at <http://www.wordspy.com/>, accessed 29 November 2011.

Yavaş, M. (2006): *Applied English Phonology*. Malden: Blackwell.

Zipf, G. (1949): *Human Behavior and the Principle of the Least Effort*. New York: Hafner.

Index

A

abbreviation 103, 105, 106, 157
absolute synonymy 37, 61
absolute tense 195
absolute transitive use 184
accusative 10, 12, 13, 17, 175, 176, 179-193, 211, 212
achievement 200-202, 219
acoustic phonetics 7
acronym 106
active voice 27, 175, 185, 206
additional naming requirement 170
additive principle of morphology 49
addressee 69, 176, 198, 204, 207, 210
adjunct 10, 11, 23, 199, 201
affix 51-60, 63, 66, 79, 80, 83, 85, 99, 100, 102, 103, 107, 117, 126, 133-144, 168, 178, 197, 207
affixation 79, 81-84, 88, 99, 100, 104, 106, 107, 122-125, 130, 131, 133, 134, 138, 141, 143, 145, 153, 167, 171, 173, 176-180, 182, 197, 206, 214
affix vocable 134
African American English 12
agent 96, 97, 132, 138, 193
Aktionsart 200
allolex 63, 65, 75, 79, 80, 90, 97-99, 103-109, 155, 173, 185
allomorph 33-39, 57, 61, 63, 98, 217
allophone 8, 33
allowordform 181-183, 208, 209
alphabetism 106
alveolar articulation 8, 35
analytic form 59, 60, 177, 184
analytic formation 176, 177, 181, 197
anisomorphic borrowing 82, 83, 85
anisomorphism 39, 45, 47
antepenult 141
anterior meaning 58
apophony 81, 83, 101, 102, 104, 106, 109, 143-145, 158, 162, 171, 176-182, 218
arbitrariness 81
arbitrary formation 81, 82, 127, 129, 130
articulatory phonetics 7
aspect 26, 27, 59, 202-206, 218
aspectualizer 202-204
atelic event 201
auxiliary verb 9, 191, 197

B

back-clipping 104
back-formation 83, 85, 87, 88, 101, 146, 168-170, 172
bahuvrihi 150, 151, 156
basic function 151
blending 83, 84, 105, 163-167, 171, 172
blocking 95-99
borrowing 82, 85, 101, 104, 106, 130-133, 145, 159
bound morph 40, 42, 44, 50, 52, 54, 55
bound morpheme 26, 27, 41

C

cardinal numeral 213, 216, 217
clipping 97, 98, 103, 104, 110, 117
coda 128, 129
cognate object 15

coinage 92, 94
combining form 25, 53, 54
comment 183
communicative rank 183
complement 10, 11, 13, 23, 30, 65, 108, 117, 118, 183, 186, 187, 190-192, 211
complementary distribution 8, 9, 32-38, 61, 63, 65, 98, 107, 108
complex sentence 12
complex word 2, 3, 7, 19-21, 26, 40, 50, 59, 66, 75, 76
compound sentence 11, 12
compounding 79, 83, 86, 99, 109-111, 145, 146, 148, 151, 152, 156, 158, 159, 162-167, 171, 172
compound of the *drum and bass*-type 150, 154, 156
compound of the *information fatigue*-type 150, 153, 156
conceptual category 4, 23
consolidation 92, 93
consonant change 143
consonant change accompanied by vowel change 143
continuative present perfect 199
continuous morpheme 27
contrastive distribution 8, 9, 21, 32, 33, 60, 63, 75
copulative compound 154, 155
creativity 90, 148
cumulative morpheme 27

D

dative shift 190
degrees of comparison 80, 174, 176, 194, 214, 216
dental sound 8
deontic modality 205
derivational affix 56, 58, 62
derivational formation 58, 134, 137, 138, 141

derivational prefix 57, 82, 130, 173
derived word 66
descriptive genitive 192
diachronic history 35, 86, 87, 89, 121, 122, 163
diachronic point of view 34, 87, 88, 122, 136
diachronic productivity 101
dictionary entry 72
differential meaning 43
direct object 186, 190, 211
direction of conversion 116, 118, 122-124
discontinuous morpheme 26
distributive event 201
ditransitive clause 186, 190
dynamic event 187, 200, 204

E

elliptical sentence 20, 21, 26
encyclopedic knowledge 5
endocentric blend 164
endocentric compound 152-155
epistemic modality 198, 205
established lexeme 95
establishment 92, 93, 95, 110
euphemism 90, 91
exocentric blend 164
exocentric compound 152-156, 163
experiencer 96
experiential co-occurrence 114, 115
experiential similarity 114
expletive infixation 107
expressive function 7
expressivity 89, 90
external phase 202

F

finite form 196
finiteness 68
first person 180, 210, 212
fixed word-order 18

fore-clipping 104
formal double-headedness 155
free distribution 29, 33, 37
free form 59, 63
free morpheme 26, 41
free variation 8, 9, 12-14, 29, 31-38, 63, 65, 98, 208
frequency criterion 123
full-idiom 66, 67, 70, 74, 76, 83, 89, 95, 113, 116, 148, 199
full-idiomatization 81, 85, 86, 124, 146
full-lexicalization 93
future tense 197, 198

G
General American 7, 12, 29, 31
genitive 10, 176, 179-181, 188-193, 218, 219
genitive meaning 192, 193
genuine compound 146
gerund 196
gram 174, 177, 190, 193, 197, 204, 206, 210, 214
grammatical apophony 177
grammatical category 2, 26, 80, 173-175, 185, 194, 200, 205, 209, 212, 213, 216
grammatical meaning 2, 9, 15, 26, 27, 57-59, 65, 75, 80, 173, 174, 181, 202, 206, 210, 211, 216, 218, 220
grammatical morpheme 26
grammeme 174, 175, 183, 184, 188, 191-193, 198, 209, 212, 216, 218

H
habitual action 59
habitual event 201
hapax legomenon 102, 103, 109
homonymic clash 96

homonymy 28, 62
hypostatization 91

I
idiomatic phrase 64, 68, 100, 105, 160, 166-168
idiomatic VP 19, 63, 64, 68, 76, 155
idiomaticity 39, 55, 58, 59, 137, 166
idiomatization of phrases and sentences 83, 84, 166
imperative 181, 206-208, 219
inchoative phase 202
indefinite gerund 196
indefinite infinitive 196
indefiniteness 68, 69
indicative mood 27, 206, 207, 209, 219
indirect object 186, 190, 211
inessive case 193
infinitive 196
infix 56, 107, 133, 144, 177
inflected form 58
inflectional affix 60
inflectional marking 40, 154, 155
input lex 79, 80, 83, 87, 97-99, 108, 137, 139, 153, 163, 168
input meaning 2
input signifier 106
instantaneous event 202
interfix 56, 133
internal stage 202, 204
interrogative sentence 9
interrogative tag 10
intransitive verb 108
irrealis mood 207
irregular wordform 179, 180
isolatability criterion 14, 16, 17, 19, 20-23, 41, 79
isomorphic affixation 82, 84
isomorphic borrowing 81, 82

isomorphic lexeme 66
isomorphism 39
iterative event 201

L

lateral sound 128
lex 64-75, 79-87, 89, 90, 93, 97-100, 102-110, 155-118, 122, 127, 129, 131, 133, 135-139, 143-146, 148, 153-157, 159-161, 163, 165, 166, 168, 170, 171, 173, 174, 177, 182, 185
lexeme 63-76, 79-103, 105-127, 129-134, 136-139, 141-146, 148-150, 153-160, 162-174, 177, 181, 182, 185, 211-214, 216, 218
lexeme family 63, 74, 76
lexeme-building mechanism 80, 101, 111, 124, 146, 166, 173, 182
lexeme-formation 79, 80
lexeme-manufacturing 81, 127, 129
lex-formation 79, 80, 103
lexical conditioning 36, 61
lexical gap 89, 90
lexical meaning 9, 21, 56-58, 63, 65, 173, 212
lexical morpheme 26
light verb 15
linguistic sign 25, 27, 29-31, 60
literal meaning 5, 6, 39, 45, 47, 48, 50, 63, 67, 73
living affix 136
loan-translation 130, 131

M

main verb 9, 18
meaning-distinguishing function 7, 8
mega-morph 48
mega-morpheme 27
mental lexicon 4, 64, 68, 70, 75, 76
meta-linguistic comment 93

metaphor 112, 114, 124, 148, 171
metaphorization 115, 119, 122, 124, 149
metonymization 120, 121, 148, 149
metonymy 112-114, 117, 121, 124, 146, 149
mid-clipping 104
middle phase 202-204, 218
minimum free form 16
modal adverb 205
modal meaning 205
modal tag 205
modalizer 205, 206
monomorphemic word 26, 40, 52, 137, 144
monotransitive clause 186, 190
mood 176, 194, 205-207, 209, 218, 219
morfoid 44-46, 50, 55, 60, 66, 67, 70, 71, 87, 90, 137
morph 25, 33-38, 40-44, 46-52, 54, 57, 59-61, 63, 64, 66, 67, 71, 83, 133, 137, 144, 174, 206, 213
morpheme 25-27, 33-44, 46, 48-50, 52-54, 57, 60- 65, 71, 134, 162, 174
morphological conversion 81, 100, 109, 115, 121, 124, 171
movement criterion 18, 19
multiplicative event 201
mutually exclusive meanings 173, 174, 185, 210, 214, 216, 218

N

necessary and sufficient conditions 4
neo-classical compound 163
neologism 92, 93, 95, 103
nominative 10, 12, 13, 175, 180, 188-193, 211, 219
nonce-formation 92, 93

non-continuative present perfect 199
non-finite form 196
non-gradable adjective 216
non-institutionalization 94-98
non-progressive aspect 27, 206
non-segmental property 83
nucleus 128, 129
numerical order 216
numerical qualification 194, 216
numerical quantity 216

O

objective meaning 192
obligatoriness 52-53, 57, 58, 159, 173, 188, 209, 213
obligatory grammatical meaning 26, 40, 56-59
oblique case 188
ongoing action 59
onset 128, 129
opacity 5, 46
opening of the vocal tract 127
optional lexical meaning 26, 56, 59, 60
ordinal numeral 176, 180, 216, 217
orthographic criterion 18, 19
orthographic modification 104, 108
output lex 79, 98, 139, 143, 144, 153

P

part-for-whole metonymy 112
part-for-part metonymy 113, 114
partial idiom 66
partial idiomaticity 45
participle I 177, 202, 203
participle II 177-179, 181, 187
partitive meaning 193
passive articulator 8
passive voice 183-185
past perfect tense 195

past simple tense 194, 212
pathological free variation 12
penultimate syllable 141
perceived similarity 112, 115, 119
perfect gerund 196
perfect infinitive 196
perfect tense 194, 195, 219
permanent event 201
permissible syllable 128
phase aspectuality 200, 201
phoneme 8, 21, 23, 27, 33, 43, 44, 65
phonological conditioning 35
phrasal verb 155, 159
phrase 9-11, 52, 63, 66, 113, 139, 155, 159, 161, 162, 184, 187, 191
place-of-articulation assimilation 35, 61, 141
plain form 207-209
plosive sound 127-129
plural number 2, 26, 44, 45, 173, 174, 210
pluralia tantum 212, 213
politeness-related context 198
polymorphemic word 26
polysemy 28, 62
popular / folk etymology 45
predicate 9-12, 15, 16, 30, 64, 65, 175, 194, 200, 206, 211
predicative 10, 12, 186
predicator 10, 11, 15, 18, 65, 117, 207, 208, 211, 212
prefix 56, 98, 99, 133, 138, 140
prepositional verb 15
present perfect tense 195
present simple tense 194, 204
present tense 2, 27, 58, 59, 173-183, 194, 197-199, 203, 204, 206, 207, 209, 212, 219
primary lex 170
progressive aspect 202
prospective aspect 203

prospective phase 202
prototype 4, 23
proverb 63, 64, 68, 75, 167
pseudo-compound 146
pseudo-concept 94
purely formal mechanism 80, 81
purely semantic mechanism 80, 111

Q

qualitative aspectuality 200
quantitative aspectuality 200, 201, 204
quasi-idiom 66, 68, 69, 70, 72, 73, 75, 84, 86, 88, 89, 95, 111, 115, 118-124, 143, 144, 151-153, 191, 193, 210
quasi-idiomatic relation 72, 144
quasi-idiomatic signified 70, 82, 146, 191
quasi-idiomatization 79, 81, 83, 117, 123, 124, 143, 148, 151, 153, 164, 168, 193, 213
quasi-linguistic unit 44-46, 51, 66, 67, 83, 133, 144, 213

R

Received Pronunciation 7, 12, 14, 29, 31
referential meaning 175, 176
register 31
regular past tense suffix 182
relative tense 195, 196, 219
resultative phase 202

S

schwa 144
second person 180, 210, 212
self-speech 210
semantic blocking 99
semantic calque 130
semantic case 193
semantic change 80, 81, 109, 111, 112, 114-116, 121, 124, 126, 145, 148, 171, 214
semantic grammeme 173, 174, 176, 194
semantic modification 2, 5, 80, 82, 83, 85, 124, 130
semantic narrowing 28, 47, 72, 113
semantic role 193
semantic widening 113
semantically more complex 2, 88, 119, 123, 124, 215
semantically more regular 58
semelfactive event 201
semi-idiom 66, 67, 69-71, 76, 77, 89, 90, 114, 149, 153, 199
semi-idiomatic relation 73, 74
signifier-sharing 181-183
simple sentence 11
simple tense 194, 195
simple word 2, 26
simultaneousness 196, 208
singular number 2, 27, 173, 174, 211
singularia tantum 212, 213
six types of genitive constructions 191
sociolinguistics 3, 30-33, 64, 65, 192
socio-pragmatic perspective 92
sonority sequencing generalization 127, 128
speech event 209, 210
static event 200
stative verb 187
stone wall-combination 117, 118
stress shift 144
stress shift accompanied by vowel change 143
stress-attracting affix 140
stress-neutral affix 140
stress-shifting affix 140

structural perspective 92
stylistically irrelevant variation 13
stylistically relevant variation 13, 14, 21, 31, 32, 35, 37, 65, 98, 208
subjectivisation of meaning 37
subjunctive 181, 206-209, 218-220
subjunctive I 207-209
subjunctive II 207-209
submorph 44-46, 50-52, 60, 66-68
suffix 56-58, 68, 79, 80, 84, 85, 88, 96, 98-103, 106, 110, 122, 125, 126, 133-142, 144, 147, 150, 153, 159, 171, 173, 178-180, 182, 184, 193, 197-199, 206, 207, 214, 217, 218
summation plural 213
suppletion 87, 103-105, 176, 180, 182, 217
suppression 184
suppressive voice 184
syllable 29, 83, 128, 140, 141
synchronic point of view 34, 88, 117, 122, 143, 145, 169
synchronic relation 86, 88
synonym 7, 29, 47
synonymic blocking 96
synonymic sets criterion 122, 123
syntactic function 9, 22, 193
syntactic grammeme 174-175, 183, 192, 194, 218
syntactics 29, 30-34, 64, 98, 99, 104, 107, 108, 115, 133-141, 185
syntactics' blocking 99

T
taboo 89, 90, 114
tactically different environment 36, 37
tactically identical environment 36
telic event 201
temporal meaning 193, 194, 196, 219
tense locus 194, 195, 198
terminal phase 202, 203
third person 178, 180, 210, 212
topic 183, 184
transitive verb 108
true etymology 67, 89, 169, 170

U
uncountable noun 213, 216
uninterruptability criterion 18, 19
unique unit 42, 46, 48, 52
unreal event 198

V
velar softening 141
vocable 63, 72-76, 86, 93, 95, 111, 119, 124, 127, 134, 213, 216, 218
vocal cords 127
voiced sound 127, 128
voiceless sound 128
voicing 127
vowel change 83, 143, 144, 177-179

W
whole-for-part metonymy 113
wordform 65, 75, 80, 123, 173-184, 188-193, 197-199, 204, 206-210, 212-220
word-formation 3, 79, 80, 110, 172
wordhood criteria 14

Z
zero morph 48-50

Textbooks in English Language and Linguistics (TELL)

Edited by Magnus Huber and Joybrato Mukherjee

Band 1   Ulrike Gut: Introduction to English Phonetics and Phonology. 2009.
Band 2   Jürgen Esser: Introduction to English Text-linguistics. 2009.
Band 3   Rolf Kreyer: Introduction to English Syntax. 2010.
Band 4   Alexander Bergs: Synchronic English Linguistics. 2012.
Band 5   Alexander Tokar: Introduction to English Morphology. 2012.

www.peterlang.de

www.ingramcontent.com/pod-product-compliance
Ingram Content Group UK Ltd.
Pitfield, Milton Keynes, MK11 3LW, UK
UKHW021324180426
11947UKWH00017B/1429